AMATEUR ASTRONOMERS — especially beginners — have long been frustrated and discouraged by conventional star charts, or planispheres, which depict the constellations arranged in a circle. While this scheme is scientifically correct, beginners have complained about the difficulty of matching up what they see in the sky with the patterns of their planispheres.

Drs. Levitt and Marshall have solved this problem brilliantly by dividing the sky into quadrants. Their charts, which look roughly like Maltese Crosses, make it possible for the user to orient himself quickly and easily in any compass direction simply by turning the chart to correspond with the section of sky he is studying. Whether one uses direct vision, binoculars or a telescope, the charts enable him to locate himself almost instantly anywhere in the northern hemisphere . . . at any hour of every night of the year.

Dr. Levitt is director emeritus of the Fels Planetarium at The Franklin Institute in Philadelphia. The late Dr. Marshall had been director of the Morehead Planetarium and professor of astronomy at the University of North Carolina, Chapel Hill. The authors were close friends for many years, and both have written other books on astronomy and space travel.

STAR MAPS
FOR
BEGINNERS

newly revised and updated

50th Anniversary Edition

BY I. M. LEVITT & ROY K. MARSHALL

A FIRESIDE BOOK
Published by Simon & Schuster
New York London Toronto Sydney Tokyo Singapore

CONTENTS

FIRESIDE
Simon & Schuster Building
Rockefeller Center
1230 Avenue of the Americas
New York, New York 10020

Copyright © 1942, 1943, 1944, 1945, 1946 © 1964, 1974 by
I. M. Levitt and Roy K. Marshall
Copyright © 1980, 1983, 1985, 1987, 1992 by I. M. Levitt
1964 Edition copyright renewed 1992

New Revised Fireside Edition 1992

FIRESIDE and colophon are registered trademarks of
Simon & Schuster Inc.

Manufactured in the United States of America

10 9 8 7 6 5 4 3

Library of Congress Cataloging-in-Publication Data
Levitt, I. M. (Israel Monroe).
    Star maps for beginners.
    "A Fireside book."
    1. Astronomy—Observers' manuals.
I. Marshall, Roy Kenneth. II. Title.
QB63.L4 1987   523.8'9   86-22899
ISBN: 0-671-79187-7

PREFACE

This book sprang from the initiative of Dr. Levitt, who persuaded the Sunday magazine editor of the Philadelphia Inquirer to publish a series of monthly star maps from June 1940 to May 1941. Dr. Marshall designed the style of the star maps, and together the authors drew the maps and provided the text that was printed each month in the newspaper.

The reaction on the part of the public was to ask why the maps could not be assembled in book form for permanent use, so a 33-page pamphlet in 8½-by-11-inch format was issued on May 1, 1942, privately published by the authors, both then at the Fels Planetarium of The Franklin Institute in Philadelphia. The text was completely rewritten and the long introductory section on the history of the constellations was added. To date over 460,000 copies have been sold.

Senior Editor Merrill Pollack of Simon & Schuster picked up a copy at The Franklin Institute, found it to be the most usable set of star maps he had tried for his own personal stargazing, and asked the authors if they would consider permitting his company to issue the book. This volume, completely revised wherever necessary, and amplified with new material such as that concerning stellar magnitudes, the use of binoculars and other amateur optical aids, and details of the sun's family, updated with the data from the Mariner, Voyager, and Pioneer spacecraft and information on celestial "hot rods," or meteors, is the result. It is sincerely hoped that in this new form and under the aegis of one of the most respected and progressive publishing houses, the work will have another long span of usefulness and popularity with those who wish to take the first steps toward knowing the sky and enjoying the delights of amateur astronomy.

I. M. LEVITT
Philadelphia
Spring 1992

EXPLANATORY NOTES

THE MAPS in this book are drawn exactly for a latitude of 40 degrees North — the parallel for Philadelphia, Indianapolis, Denver, Reno, northern Japan, Korea, Peking (China), Ankara (Turkey), northern Greece, the "foot of the boot" of Italy and Madrid (Spain). However, they will serve amply well for places as far as six or seven degrees north or south of this specific latitude (or 400 to 500 miles), thus accommodating approximately 20 per cent of the world's population.

From a position north of the fortieth parallel an observer will be able to see some stars that are beyond the northern horizon indicated on our maps, and he will be unable to see some stars that are represented near the southern horizon on our maps. Conversely, from a position south of the fortieth parallel an observer will have a more extensive view to the south and a less extensive view to the north.

Another convention we have had to adopt is that of showing the sky for the times given for the exact Standard Time meridians. Determine what Standard Time is used in your area and the longitude corresponding to it. Eastern Standard Time is based on longitude 75 degrees West; Central Standard Time on longitude 90 degrees West; Mountain Standard Time on longitude 105 degrees West; Pacific Standard Time on longitude 120 degrees West. Then, determine your own longitude, from an atlas. Take the difference in degrees between your own longitude and that of your appropriate Standard Time meridian; multiply it by 4 to convert the degrees into minutes of time. If you are east of the Standard Time meridian, subtract the minutes from the time given on the map to determine the moment to see the sky as pictured; if you are west of the Standard Time meridian, add the minutes of longitude correction to the time given.

A glance at the maps will suffice to show how to use them. The words "Looking North," "Looking East," etc., serve to orient the maps to match the sky. If you look east, the words "Looking East" should be right side up, and so on.

The charts are arranged more or less in the form of what is called a Formée Cross, except that the sides of the panels are convex instead of concave. This largely eliminates distortion; the star groups have nearly the same shapes in the sky and on the maps. If a group straddles a division between two panels of the cross, reference to a preceding or following map will show the whole outline. Many star maps designed for the beginner permit so much distortion that they defeat their purpose; practically no one who is not already acquainted with the constellations can recognize them on those maps. Too many such star charts, showing all of the sky in circular form, with the pole or zenith in the center, or half the sky as half of a circular disk, with the zenith at the top, have been circulated with too little regard for the possibility of practical use by a beginner.

Undoubtedly, many deviations from exact representation of the heavens will be spotted in the maps in this book, but they are comparatively small; moreover, because of the "open" appearance of the charts, resulting from the elimination of vast numbers of faint stars, there is never much chance for confusion.

Many people do not know that red light promotes and maintains dark adaptation — the ability of the eyes to see faint objects out-of-doors after leaving a brightly lighted house. In using the maps out of doors, a flashlight with two or three layers of red cellophane over the lens can be used, to make sure that the stars on the maps can be seen, while at the same time the stars in the sky will be plainly visible.

Some classicists may object to the mixture of Greek and Roman names in the myths. We know that the Romans borrowed the Greek myths, which in turn the Greeks had borrowed from the Phoenicians, who had borrowed them from the Babylonians, and so on. The names of the characters given here are, it is believed, the commonest ones associated with them. Many good books have been devoted to mythology *per se;* they can be found on most library shelves.

A very exhaustive book for those who wish to know the origins of the names of stars is Allen's *Star Names and Their Meanings,* which contains a wealth of information about the constellations; it has long been out of print, but it is available in many libraries. Much of the history is found in Basil Brown's *Astronomical Atlases, Maps and Charts;* this too is out of print, but can be found in some libraries.

Those who find their appetites whetted by this elementary book of maps may care to go on to a more advanced atlas, in which many more stars, as well as the Bayer and Flamsteed designations, are given. Those by Schurig-Götze and by Norton are good ones. The most modern is Antonin Becvar's de luxe *Atlas of the Heavens*, which can be purchased in a less expensive edition called *Field Atlas of the Heavens*. Prices of many such publications can be obtained from the Sky Publishing Corporation, Harvard Observatory, Cambridge 38, Mass., which also publishes a fine monthly astronomical magazine, *Sky and Telescope*.

Many amateurs (beginners, really) have asked the authors to recommend material that would be helpful in pursuing the subject. The first thought would be to consult school and public libraries; even if the selection is small and poorly chosen, the beginner can profit by reading through it. Then, with references to newer publications in such journals as *Sky and Telescope*, the better volumes being printed today can be obtained or, perhaps, recommended to the libraries. The volume of material in this Space Age is enormous.

THE HISTORY AND DEVELOPMENT OF THE CONSTELLATIONS

MODERN ASTRONOMY has become a highly specialized study, with a good knowledge of at least elementary physics and mathematics required to follow its many ramifications. We could hardly expect it to be otherwise, in a science which attempts to embrace as its field the whole of creation — the universe.

The findings of modern astronomy make fascinating reading for one who is willing to realize that no one can expect to grasp quite all of what is contained even in a so-called "popular" book, without at least some measure of concentration and connected thought. This is not peculiar to astronomy, of course; modern physics, chemistry, geology, botany — even economics and political science — are such specialized subjects that the general reader must appreciate his handicaps and must not expect to be able to grasp completely in one hurried reading what other men learn only after many years of concentrated study.

But there is one part of astronomy in which the professional astronomer has little interest, and it is in this field that the interested amateur can become as proficient as the greatest of the ancient astronomers. This is the study of the apparent face of the sky, to the end of being able to identify the star groups or constellations, and to name many of the stars. One need not be a geologist to enjoy rolling hills or soaring mountains, or a botanist to enjoy a flower; to know and enjoy the stars requires no technical knowledge, but it is an achievement of which one may well be proud. It leads to greater appreciation of great works of music, art, and literature, for the heroes who fill the sky are favorites in these other aesthetic endeavors of mankind. Today most of us read very little of the old legends of Rome and Greece, but a study of the constellations will prove an incentive to greater enjoyment of these old stories.

The sky is parceled into named areas called constellations, as our country is divided into named areas called states. It is in just this way that a modern astronomer regards the constellations — as named areas — and it is quite likely that those forgotten stargazers who originally named the constellations thought of them in the same way. Sometime in between, however, there arose a demand that a constellation named Hercules, for example, should look like Hercules, the prodigiously strong son of Jupiter. Suppose we were to insist that a state named Washington should look like our first president! Or suppose the states of Georgia, North and South Carolina, Maryland and Virginia should have their boundaries changed, to force those states to be profile portraits of a King George, a King Charles, a Queen Mary, and a Queen Elizabeth (the Virgin) of England! We should regard such a thing as at least slightly silly, yet almost everyone is under the impression that the constellations are supposed to be pictures, because they bear the names of persons, creatures, and objects.

The earliest remaining complete description of the sky as seen from Greece was written by the poet Aratus, whom we shall mention again. He stated that certain mortals, "in ages long agone," finding that it was a tedious task and not particularly helpful in identification to give a name to every star, decided to name them in groups. Then, as we might refer to "that biggest oak tree in Johnson's meadow," the early watchers of the sky might speak of "the brightest star in the constellation Auriga." How soon after the naming process the pictures were associated with the constellations we do not know, but it must have been very early.

The earliest complete representation of the heavens as they were considered at the time appears to be the famous Farnese Globe, now in the Naples Museum. Discovered in Italy, it consists of white marble, and portrays Atlas on one knee, supporting on his bowed head and shoulders the celestial sphere, which he steadies with his hands. In an excellent state of preservation, it dates from at least as early as the first century before the Christian Era. Beautifully sculptured in raised relief, in the correct positions on the sphere, are the pictures of the constellations, but images of the stars are not shown.

Similarly, the earliest manuscript map of the sky contains only the constellation figures, and not the stars. The so-called *Planisphere of Geruvigus*, included in a Roman manuscript version of Aratus, dates from the second century A.D. and is now in the British Museum. It differs from the Farnese Globe and resembles modern maps in that it represents the actual face of the sky; that is, it shows the constellations as seen from the inside of the celestial sphere, as we see them from the earth.

In the earliest map showing the constellation figures and also the stars tolerably well located, we find a return to the practice of showing the sky as it appears on the surface of the sphere, as seen from the outside. It is the work of Peter Bienewitz (Latinized as Petrus Apianus), published as a single sheet, at Ingolstadt on August 5, 1536. It is a woodcut, well executed, representing forty-eight constellations.

But it was Johann Bayer, a lawyer and amateur astronomer of Augsburg, who published (1603) the star atlas which was the prototype of a number of fine atlases prepared by later astronomers. Bayer's *Uranometria* shows the positions of about 1250 stars, with their relative brightnesses quite accurately represented, and upon the star groups are shown the constellation pictures. The fifty-one plates were exquisitely engraved on copper by Alexander Mair. Here we find again a star map showing the sky as seen from the inside, as we actually see it, and practically every map of the sky (except, of course, celestial globes) has, since that time, been drawn this way. To Bayer, too, we owe our modern method of designating most of the naked-eye stars by letters of the Greek alphabet, in each constellation. His atlas passed through several editions.

It was more than a century before the star maps of Bayer were equaled, when John Flamsteed, the British Astronomer Royal, observed the positions of the stars for a catalogue and atlas (posthumously published in 1729). The constellation figures are in some respects superior to those of Bayer, without, perhaps, the same beautiful workmanship. There were many editions of this atlas, in which the practice of numbering the stars in each constellation, in order from west to east, was established. At a later moment, we shall explain and illustrate these designation schemes of Bayer and Flamsteed.

Later star atlases were published by Doppelmayer (1742), Bevis (1750), Burritt (1851) and others, but perhaps only that of Johann Elert Bode, about 1800, need be mentioned here. Bode seems to have been the first one to draw star charts to show the skies month by month, a scheme which has been quite popular for several generations, particularly for star maps intended for the beginner. It is a similar scheme which has been followed for the charts in this book.

Besides the sculptures and maps showing the pictures over the whole sky, there have come down to us descriptions of the sky and fragmentary representations which push yet farther back our knowledge of the framers of the constellations. Originally, modern astronomers believed that the Greeks had apportioned the sky into star groups, because most of the legends connected with the figures in the sky were known to be Greek. But, with the growth of our knowledge of the civilizations of the valley of the Tigris and Euphrates rivers, there has come a realization that many of the Greek myths had a Semitic origin. The Greeks simply changed the settings and the names, and took over the plots of the legends. Might they not similarly have taken over the constellations of the Euphratean peoples?

We know that the Akkadians and Sumerians, non-Semitic forerunners of the Babylonians, had names for many of the stars, chosen particularly from the words in use in shepherding. The stars were known as the "heavenly flock"; the bright star Arcturus was called Sibzianna, the "star of the shepherds of the heavenly herds." The sun was called the "old sheep"; the planets were the "old-sheep stars." This was the kind of astronomy inherited by the Babylonians from their predecessors in the Euphratean valley.

Examination of baked-clay tablets and cylinder seals which date from 3500 to 500 B.C. gives a few clues. One of the older myths describes a battle between Marduk, city-god of Babylon, and the dragon Tiamat. On a clay cylinder seal dating from at least as early as 3000 B.C., Izhdubar (better known in English literature as Gilgamesh) is pictured kneeling on a dragon. The Greeks inherited a constellation called En Gonasin, the Kneeler, who has one foot on the head of a dragon. They were reminded of their hero Herakles (Roman Hercules) and his struggle with the Dragon which guarded the Golden Apples of the Hesperides. So Hercules and Draco are surely two very old constellations. Another is Leo, the Lion, which is shown on an ancient clay tablet, with the star Regulus marking his heart. A cuneiform synthesis of all earlier inscriptions (known as the "Creation Legend," compiled about 650 B.C., during the reign of Assurbani-pal) indicates that there were recognized thirty-six constellations, divided into three groups — northern, zodiacal, and southern.

The poems of Homer (*Iliad* and *Odyssey*, dating perhaps from the middle of the ninth century B.C., according to Herodotus) contain references to the constellations, but inasmuch as Homer was probably only the collector of the tales and ballads of earlier times, these constellations must be much older. The writings of Hesiod (*Theogonia* and *Works and Days*), about a century later, mention Arcturus, the Pleiades, the Hyades, Sirius and Orion, while Homer had referred to Ursa Major, in addition to these.

It is more than likely that the early Greeks received their astronomical lore from the Euphrateans, by way of the Phoenicians, a remarkable people who started out north of Palestine as early as 3000 B.C.; their great cities were Tyre and Sidon, but by 600 B.C. they had colonized North Africa and had founded the great city of Carthage, among others. Some of the best Greek astronomers (an instance is Thales of Miletus, about 600 B.C.) were of Phoenician descent.

Aglaosthenes (c. 650 B.C.) mentioned Aquila and Cynosura (now Ursa Minor). The early Mediterranean sailors had used what we call the Big Dipper in the northern heavens to guide them, but the Phoenicians switched to the Little Dipper, or Ursa Minor. Today we are alluding to this when we speak of something which is the center of attention as a cynosure.

Epimenides of Crete (c. 600 B.C.) wrote of Capricornus and the star Capella; Pherecydes of Athens (500–450 B.C.) told the legend of Orion and mentioned the fact that, as Orion sets, Scorpius rises; Aeschylus (526-456 B.C.) and Hellanicus of Mytilene (496–411 B.C.) tell the story of the seven Pleiades. Geminus of Rhodes relates that, in the fifth century B.C., Eustemon of Athens compiled a weather almanac in which he mentioned Orion, the Hyades, the Pleiades, Lyra, Cygnus, Aquarius, Corona, Delphinus, Pegasus, Aquila, and Canis Major as weather portents.

Eudoxus of Cnidus (c. 403–350 B.C.) appears to have been the earliest to write of constellations as such, merely for the purpose of writing a description of the sky. The title of his work was *Phaenomena*, and this title was preserved by the Cilician poet Aratus (c. 270 B.C.) when, by command of the Macedonian king Antigonus Gonatas, he put the description of the sky by Eudoxus into verse. The original work of Eudoxus has been lost.

Aratus begins with an invocation to the god Zeus and uses in the first words of the fifth verse the phrase, "For we are his children." Saint Paul, in his sermon to the Athenians (Acts 17:28), referring of course to the Supreme Being, quotes Aratus and one of his contemporaries, Cleanthes: "For in Him we live, and move, and have our being; as certain also of your own poets have said, For we are also his offspring."

In the *Phaenomena* of Aratus, forty-four constellations are named, but one of them is the small cluster we call the Pleiades and consider a part of Taurus. In addition, however, Procyon is mentioned, and this may be considered to be a recognition of Canis Minor as a separate named constellation. The star groups are placed in three regions: northern, zodiacal and southern. The zodiac, or "circle of animals," is that belt of the sky in which the sun, moon and bright planets are always to be found.

Below is the list of the constellations named, described and located relatively to each other by Aratus. The Pleiades have been omitted, and Canis Minor has been put in place of Procyon. There remain two rather unfamiliar groups, Chelae and Serpentarius, and one whose name is not to be found in up-to-date lists: Argo Navis. We shall soon take care of these.

THE CONSTELLATIONS OF ARATUS

NORTHERN

Ursa Major, Ursa Minor, Boötes, Draco, Cepheus, Cassiopeia, Andromeda, Perseus, Triangulum, Pegasus, Delphinus, Auriga, Hercules, Lyra, Cygnus, Aquila, Sagitta, Corona, Serpentarius

ZODIACAL

Aries, Taurus, Gemini, Cancer, Leo, Virgo, Chelae, Scorpius, Sagittarius, Capricornus, Aquarius, Pisces

SOUTHERN

Orion, Canis Major, Canis Minor, Eridanus, Lepus, Cetus, Argo Navis, Piscis Austrinus, Ara, Centaurus, Hydra, Crater, Corvus

Another who wrote a commentary on the *Phaenomena* of Eudoxus was Hipparchus of Bithynia, one of the greatest men of antiquity (c. 160–125 B.C.). His work in original form is not extant, but it was incorporated in a work of three centuries later, by Claudius Ptolemy. Hipparchus went further than mere description of the constellations in words; he compiled the first star catalogue, in which were listed the positions and the relative brightnesses of the stars. A century earlier, two Alexandrian astronomers, Aristillus and Timochares, had made measurements of star positions, and their work was adopted and extended by Hipparchus. It was he who inaugurated the classification of the stars by "magnitudes," the brightest stars being of the "first magnitude," the faintest visible to the naked eye being of the "sixth magnitude."

Callimachus and Eratosthenes, who were practically contemporary with Aratus, had written descriptions of the constellations, and in these a new constellation, Coma Berenices, had appeared, but Hipparchus and his successors for more than seventeen centuries seem to have overlooked it. It is recognized today as a full-fledged constellation. Hipparchus added two constellations by splitting Serpentarius into Ophiuchus and Serpens and by using some of the stars of Centaurus to form a new constellation, Lupus, the Wolf.

Hipparchus may also have been the one to introduce Equuleus and Corona Austrina, for we find them in the work of Claudius Ptolemy (c. 150 A.D.). This Alexandrian astronomer adopted practically without alteration the work of Hipparchus, and thus preserved it for us. But Ptolemy must have made some original observations, for the brightnesses of the stars are now set down as of a certain magnitude, or as a little brighter than, or a little fainter than, a certain magnitude.

For about seventeen centuries it was customary for astronomers to use the approximate brightnesses as given by Ptolemy. Then Sir William Herschel (1738–1822) and, in turn, his son Sir John (1792–1871) made extensive deep surveys of the sky, John even taking a sizable telescope to the Cape of Good Hope for a few years, to extend to the south celestial pole the work his father had done from England. These were statistical surveys, aimed at trying to discover the structure, dimensions and composition of the universe. The assumption had to be that, on the average, the stars are of the same intrinsic brightness and that their distances produce the differences in apparent brightnesses. It was essential, therefore, that the brightness scale be investigated.

About 1830, Sir John determined that the ancient magnitudes of Ptolemy were based on a geometrical, rather than an arithmetical, scale. That is, a star of magnitude 1.0 is a certain number of times as bright as a star of magnitude 2.0, which in turn is the same number of times as bright as a star of magnitude 3.0, and so on. This was verified by the physiologist Weber in 1834, and in 1859 the German psychologist G. T. Fechner put it into a general law or equation for all sensations, as Weber had suggested.

In 1856, the English astronomer N. R. Pogson carefully measured the brightnesses of many of the naked-eye stars and found that the average first-magnitude star of Hipparchus and Ptolemy was close to one hundred times as bright as the average sixth-magnitude star, so he established this as a convenient ratio; astronomers have followed it since, establishing standards and extending it to the faintest objects that can be seen or photographed through the largest telescopes.

A few stars are now considered brighter than the first magnitude, so they are called zero-magnitude stars; three are even brighter, so they have minus, or negative, magnitudes. Today, brightnesses are expressed even to the hundredth of a magnitude, although the eye can hardly distinguish differences of a tenth; magnitudes are continually being re-measured and refined with better equipment.

The scale is shown in the table of brightness ratios given here.

Magnitude Difference	Brightness Ratio
0.1	1.096
0.2	1.202
0.3	1.318
0.4	1.445
0.5	1.585
0.6	1.738
0.7	1.905
0.8	2.089
0.9	2.291
1.0	2.512
2.0	6.310
3.0	15.849
4.0	39.811
5.0	100.000

For values not in the table, we multiply the appropriate factors given. For example, a difference of 1.5 magnitudes is broken down into 1.0 and 0.5, and the ratios for these two are then multiplied: 2.512 times 1.585 equals 3.98, very nearly. A difference of 10 magnitudes is a ratio of 100 x 100, or 10,000 times in brightness.

The faintest stars observed today are of magnitude 23; Sirius, the brightest star in the sky, has an apparent visual magnitude of −1.42, or almost 24.5 magnitudes brighter than the faintest. We break this down into 5 + 5 + 5 + 5 + 4 + 0.5, and multiply the ratios. We get about 6,309,500,000, the ratio of the brightness of Sirius to that of the faintest observable star! The magnitude scale is a great boon to us, removing the necessity for using such huge numbers.

The sun's apparent magnitude is −26.8, almost 50 magnitudes brighter than the faintest observable stars, which means that the brightest object we see is almost 100,000,000,000,000,000,000 times as bright

as the faintest one. It might be of interest to note that the star Sirius is just about one magnitude below the midpoint of this scale from the sun to the faintest observable star.

Now we must return to Ptolemy, who omitted the Coma Berenices group and listed a total of forty-eight constellations. But one of them included by Aratus had its name and significance changed. Chelae had become Libra.

In the Egyptian temple of Isis at Denderah, a circular representation of the heavens has been found. At first believed to be of great antiquity, it is now known to date only from the beginning of the Christian Era, in the reign of Caesar Augustus, although it may be a "corrupted" restoration of an earlier plaque. It is a strange mixture of independently conceived Egyptian constellations and conventional Greek figures. Aratus had described Chelae as the claws of Scorpius, but in this "Circular Zodiac of Denderah" this space is occupied by a pair of scales, or a balance, and so it has remained as the constellation Libra. In this transaction, the meaning of the word zodiac, the "circle of animals," has been violated, for a balance is not a living thing, however much its delicate trembling might make it seem alive. The two brightest stars of Libra bear names to remind us of their former affiliation. They are Zubenelgenubi and Zubeneschamali — the southern and northern claws of the Scorpion.

More than fourteen centuries passed, after Ptolemy, before any new constellations were added. That any more were added at all may seem surprising, but it must be understood that the boundaries of the constellations had never been defined; the spaces between the named areas contained a few faint stars from which new constellations could yet be formed. Then, too, the Greeks could not see the many stars below their southern horizon, surrounding the south celestial pole, and the great age of exploration had to come before these stars could be observed and grouped in constellations, to complete the partition of the whole celestial sphere.

It was in Bayer's *Uranometria* (1603) that these southern stars were first shown and grouped into new constellations. From the Dutch navigator Pieter Dirchsz Keyser (or Petrus Theodori, as it was Latinized), who died in 1596, Bayer obtained a description of the sky which enabled him to fill in most of this southern part of the sphere with new constellations, some of which at least partially spilled over into the part of the sky known to the ancients but as yet unclaimed by any of the old classical constellations. The new groups are listed in the tables below.

BAYER'S NEW SOUTHERN CONSTELLATIONS

APIS (the Bee; now Musca, the Fly)
AVIS INDICA (Bird of Paradise; now Apus)
CHAMAELEON (the Chameleon)
DORADO (commonly known as the Swordfish)
GRUS (the long-necked bird, the Crane)
HYDRUS (the Water Snake, not to be confused with the classical Hydra, the Water Serpent)
INDUS (the American Indian)
PHOENIX (the mythical bird, the Phoenix)
PISCIS VOLANS (the Flying Fish; now simply Volans)
TUCANA (the bird with the strange beak, the Toucan)
TRIANGULUM AUSTRALE (the Southern Triangle)

Bayer omitted Coma Berenices, which had been revived by the Danish astronomer Tycho Brahe only a few years before the publication of the *Uranometria*, but practically all later astronomers included it. With Ptolemy's forty-eight constellations, Coma Berenices and the eleven new ones introduced by Bayer, the total became sixty.

Jacob Bartsch (c. 1599–1633), the son-in-law of the great German astronomer Johann Kepler (Tycho Brahe's greatest pupil and colleague), created three new constellations in areas in the north not claimed by others. They were Camelopardus (originally and sometimes today Camelopardalis, the Camelopard or Giraffe), Monoceros (the Unicorn) and Columba Noachi (the Dove of Noah; now simply Columba). Bartsch also stated that Isaak Habrecht, of Strassburg, had created another constellation in the south polar cap; it was Rhombus (lengthened by Lacaille to Reticulum Rhomboidalis, and now shortened to Reticulum, the Net). In 1679, Augustine Royer created Crux Australis (the Southern Cross; now Crux), which had been figured on earlier maps as a Cross, but had not yet been detached from Centaurus, whose hind legs it had formed. Our total, with these additions, stands at sixty-five constellations.

The Polish astronomer Hevelius of the city of Danzig published (posthumously, 1690) seven new groups, all in the north. They are as follows:

CANES VENATICI (the Hunting Dogs)
LACERTA (the Lizard)

LEO MINOR (the Lion Cub)
LYNX (the Lynx)
SEXTANS URANIAE (the Sextant of Urania; now simply Sextans)
SCUTUM SOBIESKII (the Shield of John Sobieski, a Polish hero-king; now simply Scutum)
VULPECULA ET ANSER (the Fox and Goose; now simply Vulpecula)

Then Nicolas Louis de Lacaille (posthumously, 1769) introduced thirteen new constellations in the southern heavens; these are given in the list below. Lacaille took the stars of Pyxis from Argo Navis, one of the ancient constellations, and tried further to introduce a new constellation Malus (the Mast, of Argo Navis), but this did not survive. But because of the great size of the old constellation of Argo, modern astronomers have partitioned it into three new groups whose names are Carina (the Keel), Puppis (the Stern), and Vela (the Sail). Argo Navis is no more.

LACAILLE'S NEW SOUTHERN CONSTELLATIONS

APPARATUS SCULPTORIS (the Sculptor's Workshop; now simply Sculptor)
FORNAX CHEMICA (the Chemist's Furnace; now simply Fornax)
HOROLOGIUM (the Clock)
CAELA SCULPTORIS (the Sculptor's Chisels; now simply Caelum)
EQUULEUS PICTORIS (the Painter's Easel; now simply Pictor)
ANTLIA PNEUMATICA (the Air pump; now simply Antlia)
OCTANS (the navigation instrument invented by John Hadley)
CIRCINUS (the Compasses)
NORMA or QUADRA EUCLIDIS (the Carpenter's Square; now simply Norma)
TELESCOPIUM (the Telescope)
MICROSCOPIUM (the Microscope)
MONS MENSAE (the Table Mountain at Cape Town; now simply Mensa)
PYXIS NAUTICA (the Mariner's Compass; now simply Pyxis)

The total number of constellations is now eighty-eight, and so it is likely to remain, for there is now no room for any more. In old atlases the constellation boundaries were drawn with an exceedingly great degree of freedom; from one author to another there were large differences. In 1928 a commission of the International Astronomical Union decided on definite boundaries for all the eighty-eight constellations, and astronomers will certainly adhere to these from now on. The complete modern list is given in the table.

Today the ancient figures are almost forgotten; the constellations are considered to be quite arbitrary areas of the sky, for the purpose of convenience only. As we can locate a city fairly accurately by naming the state in which it is found and describing its location within the state, so we can designate a

THE MODERN LIST OF CONSTELLATIONS
[*as standardized by the International Astronomical Union in 1928*]

ANDROMEDA	Princess of Ethiopia
°ANTLIA	The Air Pump
°APUS	The Bird of Paradise
AQUARIUS	The Water Bearer
AQUILA	The Eagle
°ARA	The Altar
ARIES	The Ram
AURIGA	The Charioteer
BOÖTES	The Bear Driver
°CAELUM	The Sculptor's Chisel
CAMELOPARDUS	The Giraffe
CANCER	The Crab
CANES VENATICI	The Hunting Dogs
CANIS MAJOR	The Greater Dog
CANIS MINOR	The Lesser Dog
CAPRICORNUS	The Sea Goat
°CARINA	The Keel (of Argo Navis)
CASSIOPEIA	Queen of Ethiopia
°CENTAURUS	The Centaur
CEPHEUS	King of Ethiopia
CETUS	The Sea Monster
°CHAMAELEON	The Chameleon
°CIRCINUS	The Compasses
COLUMBA	The Dove (of Noah)
COMA BERENICES	Berenice's Hair
°CORONA AUSTRINA	The Southern Crown
CORONA BOREALIS	The Northern Crown
CORVUS	The Crow (or Raven)
CRATER	The Cup
°CRUX	The Southern Cross
CYGNUS	The Swan
DELPHINUS	The Dolphin
°DORADO	The Swordfish
DRACO	The Dragon
EQUULEUS	The Foal
ERIDANUS	The River
°FORNAX	The Laboratory Furnace
GEMINI	The Twins
GRUS	The Crane
HERCULES	Hercules
°HOROLOGIUM	The Clock
HYDRA	The Water Serpent
°HYDRUS	The Water Snake
°INDUS	The American Indian
LACERTA	The Lizard
LEO	The Lion
LEO MINOR	The Lion Cub
LEPUS	The Hare
LIBRA	The Beam Balance
LUPUS	The Wolf
LYNX	The Lynx
LYRA	The Lyre
°MENSA	The Table Mountain
MICROSCOPIUM	The Microscope
MONOCEROS	The Unicorn
°MUSCA	The Fly
°NORMA	The Carpenter's Square
°OCTANS	The Octant
OPHIUCHUS	The Serpent Holder
ORION	The Great Hunter
°PAVO	The Peacock
PEGASUS	The Winged Horse
PERSEUS	The Hero
°PHOENIX	The Phoenix
°PICTOR	The Painter's Easel
PISCES	The Fishes
PISCIS AUSTRINUS	The Southern Fish
PUPPIS	The Stern (of Argo Navis)
PYXIS	The Compass Box (of Argo)
°RETICULUM	The Net
SAGITTA	The Arrow
SAGITTARIUS	The Archer
SCORPIUS	The Scorpion
°SCULPTOR	The Sculptor's Workshop
SCUTUM . . .	The Shield (of John Sobieski)
SERPENS	The Serpent
SEXTANS	The Sextant
TAURUS	The Bull
°TELESCOPIUM	The Telescope
TRIANGULUM	The Triangle
°TRIANGULUM AUSTRALE .	The Southern Triangle
°TUCANA	The Toucan
URSA MAJOR	The Greater Bear
URSA MINOR	The Lesser Bear
°VELA	The Sail (of Argo Navis)
VIRGO	The Maiden
°VOLANS	The Flying Fish
VULPECULA	The Fox

°*Seen either poorly or not at all from the United States.*

star by describing its position within a constellation, or by adding something descriptive of its color or brightness.

In Bayer's *Uranometria* Greek letters were used to designate the individual stars in each constellation. For example, the star which the Arabs had indicated as Ibt-al-Jauza, the "Armpit of the Central One," and whose name had later been corrupted to Betelgeuse, was designated as α Orionis, or "alpha of Orion." The bright star Rigel, in the same constellation, was called β Orionis (beta of Orion). In this scheme, the Latin genitive, or possessive, form of the constellation name might be considered the "family name" of the star, and a particular Greek letter the "given name." In general, the Greek letters were assigned in the order of the brightnesses of the stars in the constellation: α is the brightest, β second brightest, γ the third, and so on. There are, however, a number of exceptions. In some of these it can be seen that the order is the more or less random listing of Ptolemy. In the catalogue compiled by that writer from the work of Hipparchus and others, the stars in each constellation are grouped into classes of brightnesses, with no particular arrangement in each class. In Bayer's designations, a few exceptions to the current conventional order may be the result of actual changes in brightness since his day.

In Flamsteed's atlas of 1729, the stars are designated by numbers. A star which is westernmost in a given constellation is designated number 1; the one which is next most westerly is number 2; finally the easternmost star in the constellation bears the highest number. For example, the star Betelgeuse is 58 Orionis, because it lies well toward the east in the constellation, while Rigel, near the western edge of Orion, is 19 Orionis. This scheme, as well as that of Bayer, finds general acceptance and use today; in addition, we still use some of the Arabic proper names.

The gap of more than fourteen centuries between Ptolemy and Bayer was marked by little activity in astronomy in Europe, but the Arabs preserved Ptolemy's work and added a little to the progress of the subject. They were particularly active in the naming of individual stars. Most of the names are of Arabic origin, usually very bad corruptions of the original descriptions. Betelgeuse is a good example. For most of these names there are several variations. Only scholars of Arabic can advise us concerning the pronunciations of the uncorrupted names; someday we may be able to decide on the pronunciations of the others. Even for the purely Latin constellation names, there is no agreement as to pronunciation. In the body of the text describing each map, many star names are given, in addition to several included in the maps themselves.

MOTIONS IN THE SKY

ALMOST EVERYONE knows that the earth rotates on its axis, and that due to this spinning we are carried eastward beneath the sun, thus causing the apparent motion of the sun across the sky each day. This must produce the same kind of apparent motion for any object, so the stars also appear to rise, arch across the sky, and set in the west.

If an observer were located precisely at the earth's north pole, the point exactly overhead — the north celestial pole — would appear to stand still, and all the stars would appear to describe circles about that point, each twenty-four hours. But we are about halfway from the north pole to the equator, so the north celestial pole stands about halfway up in the northern sky, instead of overhead. As the earth turns once each twenty-four hours, the stars all appear to move in circles about the north celestial pole as a center. The so-called North Star is quite close to the north pole of the sky, and we may regard this star as the center about which the sky turns each twenty-four hours.

However, in addition to the rotation on its axis, the earth has another very important motion; each year the earth makes one complete trip about the sun, in a path called its orbit. We see the sun in a different direction each day, because we stand in a different direction from the sun each day. Or we might think of it in this way: Suppose on one day the sun is in line with a certain star, as seen from the earth. On this day, the sun and the star will appear to rise together. On the next day, the earth will be a little farther along in its orbit, and the sun will appear to be a little to the east of the star. The star will rise first, and the sun will lag behind by almost four minutes. On the next day, the sun will lag almost eight minutes behind the star, and so on. Because the time we use is based upon the sun, and not upon the stars, we usually think of the stars as rising earlier each day, and slipping westward almost four minutes each day. In a month, this amounts to two hours, so each month we look for the same stars in the same places in the sky two hours earlier.

The North Star may be considered the center about which the stars describe the circles mentioned above. A star only a short distance from the North Star will trace out a small circle, one farther away will trace out a larger one, and so on. The small circles will be completely above the horizon; stars close to the North Star never set, while those far from it rise and set. There are, below our southern horizon, stars which never rise for us.

A study of the maps, in conjunction with what has been given in this section, may help the student to understand the behavior of the sky through the hours of the night and the days of the year.

THE SYNOPTIC TABLE

While each chart is marked for particular months and certain times, each chart can be used in another month at another time. Herewith is given a table to guide an observer in a selection of the correct chart for a given hour at a given time of year. For example, for July 16, from 8 to 10 P.M., the proper chart is No. 7; on the same night from 10 to 12, use chart No. 8. In the date column find the date which is nearest the exact date of observation; then look in the column headed by the time of observation. When a blank space is found, use the map indicated on either side of it.

HOW TO CHOOSE THE PROPER MAP FOR USE AT ANY TIME

Month	Date	EVENING HOURS							MORNING HOURS					
		6	7	8	9	10	11	12	1	2	3	4	5	6
Jan.	1	11		12		1		2		3		4		5
	16		12		1		2		3		4		5	
Feb.	1	12		1		2		3		4		5		6
	16		1		2		3		4		5		6	
Mar.	1	1		2		3		4		5		6		7
	16		2		3		4		5		6		7	
Apr.	1			3		4		5		6		7		
	16		3		4		5		6		7		8	
May	1			4		5		6		7		8		
	16		4		5		6		7		8		9	
June	1			5		6		7		8		9		
	16			6		7		8		9				
July	1			6		7		8		9		10		
	16		6		7		8		9		10		11	
Aug.	1			7		8		9		10		11		
	16		7		8		9		10		11		12	
Sep.	1			8		9		10		11		12		
	16		8		9		10		11		12		1	
Oct.	1	8		9		10		11		12		1		2
	16		9		10		11		12		1		2	
Nov.	1	9		10		11		12		1		2		3
	16		10		11		12		1		2		3	
Dec.	1	10		11		12		1		2		3		4
	16		11		12		1		2		3		4	

THE PLANETS AS BRIGHT STARS

IN ADDITION TO the so-called "fixed stars" shown on the maps, there are other starlike objects to be found in the sky. These are the planets, or "wandering stars," which look like bright stars, but change their positions in the constellations of the zodiac in a way that baffled the ancients.

The earth is one of the planets, revolving about the sun as do the others. As we watch those moving bodies from the moving earth, they appear to move sometimes fast, sometimes slowly, sometimes backward and sometimes forward.

The five planets visible to the naked eye are Mercury, Venus, Mars, Jupiter and Saturn. Of these, only the last four need be considered, for it is very difficult to find Mercury without very specific information. These four are quite bright and look like stars. But Mars is of a red hue, Venus is seen only for a while after sunset or before sunrise, and Jupiter and Saturn move so slowly that if one is at all acquainted with the constellations he can hardly fail to recognize them. Their approximate positions for the years 1992 to 1997 are given in the tables on the following page.

The way to use these tables is fairly simple. In the appropriate table find the date that is nearest to that of the observation; on that line, under the proper heading, appears the name of the constellation in which the planet in question appears. If the constellation appears on the map, the planet should be easily recognizable as a bright starlike object not shown on the map. If the planet is Jupiter or Venus, it will be brighter than any star. If it is Saturn, it will be about as bright as the brightest stars. If it is Mars, it will be very bright when it rises at about sunset, but only moderately bright when it rises several hours before or after sunset.

The blank spaces in the columns for Venus indicate that the planet is too near the sun to be seen. This happens to the other planets too, but only for relatively brief periods, so there are no blank spaces in their columns.

Many cities today have planetarium installations, where the motions of the planets are visually represented in compressed time. Also, at the book counters of many of these institutions a great many publications are available from which the amateur can make a selection not only to help him in his naked-eye study of the sky but also to introduce him to other fields of astronomy.

Almanacs are available which give information on the day-to-day configurations of the planets and the moon.

WHERE TO FIND THE PLANETS

1992

		Venus	Mars	Jupiter	Saturn
Aug.	1	Leo	Taurus	Leo	Capricornus
	16	Leo	Taurus	Leo	Capricornus
Sept.	1	Virgo	Taurus	Leo	Capricornus
	16	Virgo	Gemini	Virgo	Capricornus
Oct.	1	Libra	Gemini	Virgo	Capricornus
	16	Libra	Gemini	Virgo	Capricornus
Nov.	1	Scorpius	Gemini	Virgo	Capricornus
	16	Sagittarius	Cancer	Virgo	Capricornus
Dec.	1	Sagittarius	Cancer	Virgo	Capricornus
	16	Capricornus	Gemini	Virgo	Capricornus

1993

		Venus	Mars	Jupiter	Saturn
Jan.	1	Capricornus	Gemini	Virgo	Capricornus
	16	Aquarius	Gemini	Virgo	Capricornus
Feb.	1	Pisces	Gemini	Virgo	Capricornus
	16	Pisces	Gemini	Virgo	Capricornus
Mar.	1	Pisces	Gemini	Virgo	Capricornus
	16	Pisces	Gemini	Virgo	Capricornus
Apr.	1	Pisces	Gemini	Virgo	Capricornus
	16	Pisces	Gemini	Virgo	Aquarius
May	1	Pisces	Cancer	Virgo	Aquarius
	16	Pisces	Cancer	Virgo	Aquarius
June	1	Pisces	Leo	Virgo	Aquarius
	16	Aries	Leo	Virgo	Aquarius
July	1	Taurus	Leo	Virgo	Aquarius
	16	Taurus	Leo	Virgo	Aquarius
Aug.	1	Taurus	Leo	Virgo	Aquarius
	16	Gemini	Virgo	Virgo	Aquarius
Sept.	1	Cancer	Virgo	Virgo	Capricornus
	16	Leo	Virgo	Virgo	Capricornus
Oct.	1	Leo	Virgo	Virgo	Capricornus
	16	Virgo	Libra	Virgo	Capricornus
Nov.	1	Virgo	Libra	Virgo	Capricornus
	16	Libra	Scorpius	Virgo	Capricornus
Dec.	1	Libra	Scorpius	Virgo	Capricornus
	16	Scorpius	Sagittarius	Virgo	Capricornus

1994

		Venus	Mars	Jupiter	Saturn
Jan.	1	Sagittarius	Sagittarius	Libra	Capricornus
	16	Sagittarius	Sagittarius	Libra	Aquarius
Feb.	1	Capricornus	Capricornus	Libra	Aquarius
	16	Aquarius	Capicornus	Libra	Aquarius
Mar.	1	Aquarius	Capricornus	Libra	Aquarius
	16	Pisces	Aquarius	Libra	Aquarius
Apr.	1	Aries	Aquarius	Libra	Aquarius
	16	Aries	Pisces	Libra	Aquarius
May	1	Taurus	Pisces	Libra	Aquarius
	16	Taurus	Pisces	Libra	Aquarius
June	1	Gemini	Aries	Virgo	Aquarius
	16	Cancer	Aries	Virgo	Aquarius
July	1	Leo	Taurus	Virgo	Aquarius
	16	Leo	Taurus	Virgo	Aquarius
Aug.	1	Virgo	Taurus	Virgo	Aquarius
	16	Virgo	Gemini	Virgo	Aquarius
Sept.	1	Virgo	Gemini	Libra	Aquarius
	16	Libra	Gemini	Libra	Aquarius
Oct.	1	Libra	Gemini	Libra	Aquarius
	16	Libra	Cancer	Libra	Aquarius
Nov.	1	Libra	Cancer	Libra	Aquarius
	16	Virgo	Leo	Libra	Aquarius
Dec.	1	Virgo	Leo	Libra	Aquarius
	16	Libra	Leo	Scorpius	Aquarius

1995

		Venus	Mars	Jupiter	Saturn
Jan.	1	Libra	Leo	Scorpius	Aquarius
	16	Scorpius	Leo	Scorpius	Aquarius
Feb.	1	Sagittarius	Leo	Scorpius	Aquarius
	16	Sagittarius	Leo	Scorpius	Aquarius
Mar.	1	Capricornus	Cancer	Scorpius	Aquarius
	16	Capricornus	Cancer	Scorpius	Aquarius
Apr.	1	Aquarius	Cancer	Scorpius	Aquarius
	16	Pisces	Cancer	Scorpius	Aquarius
May	1	Pisces	Leo	Scorpius	Pisces
	16	Aries	Leo	Scorpius	Pisces
June	1	Aries	Leo	Scorpius	Pisces
	16	Taurus	Leo	Scorpius	Pisces
July	1	Taurus	Leo	Scorpius	Pisces
	16	Gemini	Virgo	Scorpius	Pisces
Aug.	1	Cancer	Virgo	Scorpius	Pisces
	16	Leo	Virgo	Scorpius	Pisces
Sept.	1	Leo	Virgo	Scorpius	Pisces
	16	Virgo	Virgo	Scorpius	Pisces
Oct.	1	Virgo	Libra	Scorpius	Aquarius
	16	Libra	Libra	Scorpius	Aquarius
Nov.	1	Libra	Scorpius	Scorpius	Aquarius
	16	Scorpius	Scorpius	Scorpius	Aquarius
Dec.	1	Sagittarius	Sagittarius	Scorpius	Aquarius
	16	Sagittarius	Sagittarius	Scorpius	Aquarius

1996

		Venus	Mars	Jupiter	Saturn
Jan.	1	Capricornus	Sagittarius	Sagittarius	Aquarius
	16	Aquarius	Sagittarius	Sagittarius	Aquarius
Feb.	1	Aquarius	Capricornus	Sagittarius	Aquarius
	16	Pisces	Aquarius	Sagittarius	Pisces
Mar.	1	Pisces	Aquarius	Sagittarius	Pisces
	16	Aries	Pisces	Sagittarius	Pisces
Apr.	1	Taurus	Pisces	Sagittarius	Pisces
	16	Taurus	Pisces	Sagittarius	Pisces
May	1	Taurus	Aries	Sagittarius	Pisces
	16	Taurus	Aries	Sagittarius	Pisces
June	1	Taurus	Aries	Sagittarius	Pisces
	16	Taurus	Taurus	Sagittarius	Pisces
July	1	Taurus	Taurus	Sagittarius	Pisces
	16	Taurus	Taurus	Sagittarius	Pisces
Aug.	1	Taurus	Gemini	Sagittarius	Pisces
	16	Gemini	Gemini	Sagittarius	Pisces
Sept.	1	Gemini	Gemini	Sagittarius	Pisces
	16	Cancer	Cancer	Sagittarius	Pisces
Oct.	1	Leo	Cancer	Sagittarius	Pisces
	16	Leo	Leo	Sagittarius	Pisces
Nov.	1	Virgo	Leo	Sagittarius	Pisces
	16	Virgo	Leo	Sagittarius	Pisces
Dec.	1	Libra	Leo	Sagittarius	Pisces
	16	Libra	Leo	Sagittarius	Pisces

1997

		Venus	Mars	Jupiter	Saturn
Jan.	1	Scorpius	Virgo	Sagittarius	Pisces
	16	Sagittarius	Virgo	Sagittarius	Pisces
Feb.	1	Sagittarius	Virgo	Capricornus	Pisces
	16	Capricornus	Virgo	Capricornus	Pisces
Mar.	1	Aquarius	Virgo	Capricornus	Pisces
	16	Pisces	Virgo	Capricornus	Pisces
Apr.	1	Pisces	Leo	Capricornus	Pisces
	16	Aries	Leo	Capricornus	Pisces
May	1	Aries	Leo	Capricornus	Pisces
	16	Taurus	Leo	Capricornus	Pisces
June	1	Taurus	Virgo	Capricornus	Pisces
	16	Gemini	Virgo	Capricornus	Pisces
July	1	Cancer	Virgo	Capricornus	Pisces
	16	Leo	Virgo	Capricornus	Pisces
Aug.	1	Leo	Virgo	Capricornus	Pisces
	16	Virgo	Virgo	Capricornus	Pisces
Sept.	1	Virgo	Libra	Capricornus	Pisces
	16	Virgo	Libra	Capricornus	Pisces
Oct.	1	Libra	Scorpius	Capricornus	Pisces
	16	Scorpius	Scorpius	Capricornus	Pisces
Nov.	1	Scorpius	Scorpius	Capricornus	Pisces
	16	Sagittarius	Sagittarius	Capricornus	Pisces
Dec.	1	Sagittarius	Sagittarius	Capricornus	Pisces
	16	Capricornus	Sagittarius	Capricornus	Pisces

THE JANUARY SKY

THE YEAR opens well, with the finest display of bright stars to be seen at any time. Almost due south we find Orion, the best of all star groups, marked by the unmistakable line of three stars, so evenly matched and so nicely spaced, forming the belt of the Great Hunter. Orion faces us so the red star Betelgeuse marks his right shoulder and Rigel his upraised left foot. The bright star Bellatrix, to the right of Betelgeuse, is Orion's left shoulder, and the star to the left of Rigel is Saiph, which at the same time marks the Hunter's right knee and the blade of his sword. The hilt of the sword is the little group of faint stars below the middle of the belt. To the right of Bellatrix there is a curving line of faint stars indicating the shield of the lion's hide, on Orion's upraised left arm; above Betelgeuse is a group of stars to mark his upraised right arm and club. The small triangle of faint stars above Betelgeuse and Bellatrix is his head; the Arabs called this Al Hakah, the White Spot.

Orion's gaze is directed to the right and upward, where we find the V-shaped group called the Hyades, marking the face of Taurus, the Bull. The orange star Aldebaran is the right eye of Taurus, the vertex of the V is the tip of his nose, and the star at the end of the upper arm of the V is his left eye. His head is lowered, and the tips of his horns are the two stars above Orion. The cluster called the Pleiades is in the shoulder of Taurus.

Only the head, shoulders, and forelegs of Taurus appear in the sky, for the story is that this is the white bull into which Jupiter transformed himself when he swam away from Phoenicia to Crete with the princess Europa on his back. As he swam, his hind quarters were beneath the waves, so they are not shown in the sky!

The Pleiades were daughters of Atlas, and nymphs of Diana's train. Orion saw them and pursued them, but Jupiter came to the rescue by transforming them first into pigeons, then into the stars we see in the sky in such position that Orion still pursues them. There were seven daughters, but one of them, Electra, is said to have left her place so she might not have to see the destruction of Troy, which was founded by her son Dardanus. We see only six, unless we have exceptionally good eyes, and the conditions are far above average. Then as many as nine or eleven may be seen. Even a small telescope will show scores of stars, and long-exposure photographs reveal that Tennyson was relaying scientific information when he wrote in "Locksley Hall"

Many a night I saw the Pleiads,
rising thro' the mellow shade,
Glitter like a swarm of fire-flies
tangled in a silver braid.

Most of the Pleiades cluster is involved in a nebula — "a silver braid."

Most of the material of the universe is organized into stars. Here and there, however, there are clouds of gas called nebulae. Today we recognize two fundamental kinds of nebulae — bright and dark. We know that the bright nebulae shine because they involve stars. In some instances the gas simply reflects the starlight, and so it is with the Pleiades. But the finest of all nebulae, so bright that it can be seen with the naked eye, is of the type which shines by a fluorescent process: the gas absorbs the stellar radiation, digests it, and reradiates it in another form. There are many bright nebulae of both types, but this finest example of the "fluorescent" type is also in our January sky, in the sword hilt of Orion, beneath the belt. On moonless nights it is easy to see that this is a hazy region, and with a telescope the magnificent greenish nebula is revealed, involving several faint stars.

Returning to the stars, we find in our winter sky the brightest of all of them. It is Sirius, down to the left of Orion, in the lower corner of an almost equilateral triangle which has as its other two corners the stars Betelgeuse in Orion and Procyon in Canis Minor. This large triangle is shown to better advantage on the next map. Sirius, the Dog Star (so called because it is in Canis Major, the Greater Dog), is really twenty-seven times as bright as our sun, but because it is more than 500,000 times as far away as the sun it appears to be much fainter.

The star Capella, almost overhead, is intrinsically five times as bright as Sirius, but it is also five times as far away and so appears fainter. The name of this star means "little she-goat," and the triangle of stars nearby marks the Kids, or baby goats. They are pictured on the left arm of Auriga, the Charioteer. He was more than a chariot driver; this figure is believed to memorialize the inventor of the chariot. Through Auriga runs the thin winter Milky Way. Southward it flows through the triangle Procyon-Betelgeuse-Sirius; northward of Capella the Milky Way can be traced more easily through Perseus and Cassiopeia to the northwestern horizon.

From Pisces high in the west, through Aries, Tau-

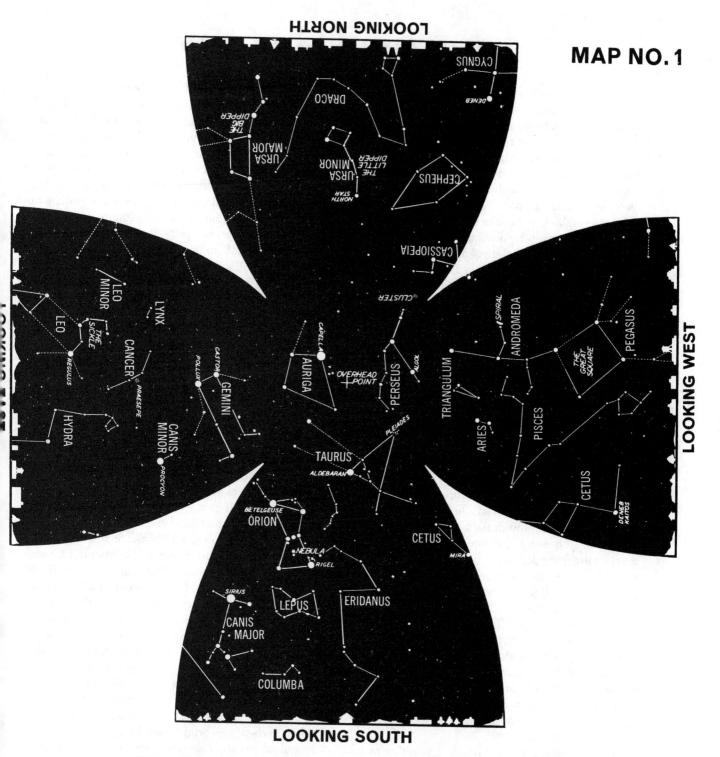

LOOKING NORTH

LOOKING SOUTH

LOOKING WEST

**This map represents the sky
at the following standard times**

**JANUARY 1 at 10 p.m.
JANUARY 16 at   9 p.m.
FEBRUARY 1 at   8 p.m.**

rus, Gemini, Cancer, and Leo in the east, we can follow the apparent path of the sun across the sky. These are six of the twelve constellations of the band of the sky known as the zodiac, in which the sun, moon and planets are to be found. If an interloper star is found in one of these constellations, the table of the planet positions given elsewhere in this book should be consulted, so it can be identified.

THE FEBRUARY SKY

THE BRILLIANT winter stars are now at their best, in the south and overhead. Low in the south are the constellations which can be seen better from the southernmost parts of the United States, but Eridanus, sometimes associated with the River Po, appears as a string of moderate stars beginning near Rigel in Orion, and is well marked, even as far north as New York or Chicago. From that far north, however, the brightest star of the constellation, Achernar, is about ten degrees below the horizon.

Above to the left of Betelgeuse (or Betelgeux, as it is alternatively written) are the feet of Gemini, the twin sons of Leda, whom Jupiter wooed in the guise of a swan. They were the half brothers of Helen of Troy; their names were Castor and Pollux. They were particularly regarded as the patron deities of sailors, and many vessels have borne their names, either separately or paired. They were highly regarded also by Roman horse soldiers, who swore by them; our expression "By Jiminy" is an oath, "By the Twins."

In the northeast the Big Dipper hangs with its handle down. The two stars in the front of the cup of the dipper are known as The Pointers, because they point very nearly to the North Star, at the end of the handle of the Little Dipper. A line drawn through the North Star, from the handle of the Big Dipper, will pass through the zigzag line of stars of Cassiopeia, a queen of Ethiopia about whose vanity a very interesting legend will be told in connection with the December map. The constellations Cepheus, Perseus, Pegasus, Andromeda and Cetus are included in the legend.

While we shall not tell the legend at this point, it is not too early to pick out some rather special objects among these constellations. Between the principal stars of Perseus and Cassiopeia is a star cluster, appearing as a bright spot in the Milky Way. Through binoculars or a low-power telescope this is a pleasing object, consisting of two swarms of stars practically in contact.

Another aggregation of stars of quite different nature and significance is in Andromeda. It is faintly visible to the naked eye as a hazy spot; the world's largest telescopes show it as a spiral formation of stars, but only by means of photographic film which can accumulate light over a long time exposure. There are millions of these great "galaxies" of stars within reach of large telescopes, and today we know them to be the units into which the matter of the universe is arranged.

We look out into a clear night sky to see stars and a few nebulae and star clusters; with powerful telescopes, millions of other stars and hundreds of clusters and nebulae are visible. All these objects are in our neighborhood, as cosmic distances go, and belong to our galaxy, or stellar aggregation. We might think of it as our "island" in the ocean of emptiness which is the whole universe. There are millions of other "islands" known, and undoubtedly countless millions of others not yet seen. Each galaxy consists of thousands of millions of stars, and each is some one or two million light-years from its neighbors. We live in one such galaxy, the Milky Way system; the spiral in Andromeda is another, about 2,200,000 light-years distant.

A light-year is a unit of distance — the distance a ray of light can span in a year, at the speed of 186,380 miles per second. A light-year is the equivalent of approximately 6,000,000,000,000 miles! The distance of the sun from the earth is about 500 light-seconds, or 500 times 186,380 miles; the distance of Sirius, one of the nearest stars, is 8.7 light-years. The largest telescopes photograph galaxies at distances of 6,000,000,000 light-years!

In Perseus is an interesting star for occasional watching. Its name is Algol, a corruption of *al-ghul*, "the demon." It is a "variable star," or one which changes in apparent brightness. Thousands of variable stars are known, and some of them we know are actually changing in size and temperature, hence are intrinsically variable. But Algol belongs to the class which we might call "accidental" variables. Algol is in reality a pair of stars revolving about their common center of gravity, like the earth-moon system. The two stars are much too far from us and much too close to each other to be separately seen even with a powerful telescope, so we see the blended light of the two as a single star.

Because the plane of the motion of the stars is edge-on to us, the two stars alternately eclipse each other. When the bright star hides the faint star, casual observation does not reveal any diminution in the light. But when the fainter star stands between us and the bright one, the result is a loss of three quarters of the combined light of the pair. Because of the reason for this change, such objects are called eclipsing variables. Observe Algol each clear night, and eventually it will be caught faint. It requires about nine hours to go through its dimming and subsequent brightening. Its period or interval between fadings is about two days and twenty-one hours.

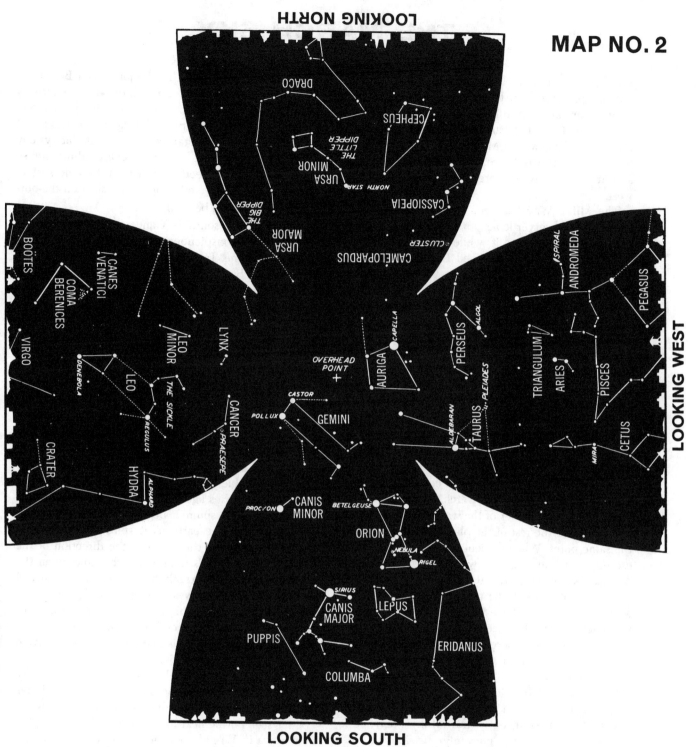

**This map represents the sky
at the following standard times**

FEBRUARY   1 at 10 p.m.
FEBRUARY 15 at   9 p.m.
MARCH   1 at   8 p.m.

Two other such bright eclipsing variables are in our February sky, but their changes are not likely to be noticed, for their periods are quite long. The star in the north corner of the little triangle in Auriga, near Capella, is Epsilon Aurigae, with a period of twenty-seven years; the star in the southwest corner of the triangle is Zeta Aurigae, with a period of about two and a half years.

THE MARCH SKY

EARLY IN MARCH we begin to see the stars which will be prominent in the skies of spring, yet the winter stars are still to be seen in the west. Before the year is too far advanced, we should pay some attention to the winter Milky Way, which now extends almost precisely from the north to the south, and arches somewhat more than halfway up in the west.

The Milky Way has been purposely left off the maps. If an observer looks at the sky from a city he will not be able to see it, while if he does his stargazing in a place far removed from city lights he will not need to have the Milky Way traced on a map. It passes through Cepheus, Cassiopeia, Perseus, Auriga, the feet of Gemini, then between Procyon and Sirius, and below the horizon after passing through Puppis. In Cassiopeia, Perseus and Auriga, it is moderately strong, but it grows weaker as it continues southward.

Remember that we are imbedded in the great system of stars which is called our galaxy. Shaped somewhat like a pocket watch, the galaxy is about 100,000 light-years in diameter, perhaps 15,000 light-years thick, and contains approximately 100,000,000,000 stars. The sun is one of these stars, located about 30,000 light-years from the center. It is, so to speak, inside the face of the watch, a little more than halfway from the center of the watch to the rim; and the earth and the rest of the planets are naturally at the same point. When we look out toward the nearest edge of the "watch," we are looking toward Auriga and Perseus; toward the center of the watch is in the direction of the tail of the Scorpion (see Maps No. 6 to 9). If we would look up, on a line perpendicular to the face of the watch, we would look toward Coma Berenices, which in March is well up in the eastern sky at the hours for which the map is drawn.

Where we look by a long path through the watch-shaped mass, we see a great many stars. This produces the appearance of the Milky Way. But because the stars are unevenly distributed, and because mingled with the stars are great clouds of obscuring dust and gases, the Milky Way is ragged and irregular. The clouds of "dark nebulae" hide from us many of the stars, and prevent our seeing out into the rest of the universe, in the direction of the Milky Way, so we find practically none of the other galaxies in or very near the directions in which the stars of our own galaxy are most densely distributed.

In the directions of Ursa Major, Coma Berenices, Boötes, Virgo and Leo, we look through a relatively thin layer of this interstellar "fog"; in these directions we can photograph with large telescopes about as many galaxies as stars of our own stellar system. As a matter of fact, our own galaxy thins out so quickly in the direction of Coma Berenices that, with an inexpensive telescope, faint stars on the outermost fringe can be seen.

High in the south is Cancer, the Crab; it is a small and inconspicuous constellation, but it is an interesting one. In it we find the cluster Praesepe, mentioned by Aratus as a weather portent. When the eye can scarcely see the cluster, while faint isolated stars nearby can be seen, there is a possibility of rain, for even a very small amount of excess water vapor in the air is enough to obscure Praesepe. There has been some disagreement concerning the meaning of the name of this cluster. Some call it the "Beehive," while others say it is the "Manger." Surely the latter is correct, for the two faint stars which flank Praesepe are respectively Asellus borealis and Asellus australis — the Northern Ass and the Southern Ass — and surely asses are not so foolish as to feed from a beehive! The literal meaning of the name is "stable," or, more specifically, "manger."

Many centuries ago the sun stood in the direction of Cancer when summer began. Now, due to a slow "wobbling" of the earth as it spins, when summer begins on June 22 the sun is in the direction of the feet of Gemini. This date marks the time when the sun, which has apparently been traveling northward for six months, turns and begins to go the other way. This used to occur when the sun stood in Cancer, and we know the crab can travel equally well one way or another. This may account for the name of this small constellation.

The star Arcturus can be found in the heavens by extending the curve of the handle of the Big Dipper outward, toward the south. This can perhaps be better seen on the next map, where it is easy to pick out Spica in Virgo by extending this curve an equal distance beyond Arcturus. These two bright stars of spring and summer are very different in color.

Differences in color indicate differences in temperature. A cool star is only "red hot"; a much hotter star is "white hot." Yellow stars like the sun or Capella are moderately hot. The word "cool" is only relative, however; the red star Betelgeuse has an effective temperature of approximately 5500° Fahrenheit. Capella and the sun are at about 10,000°,

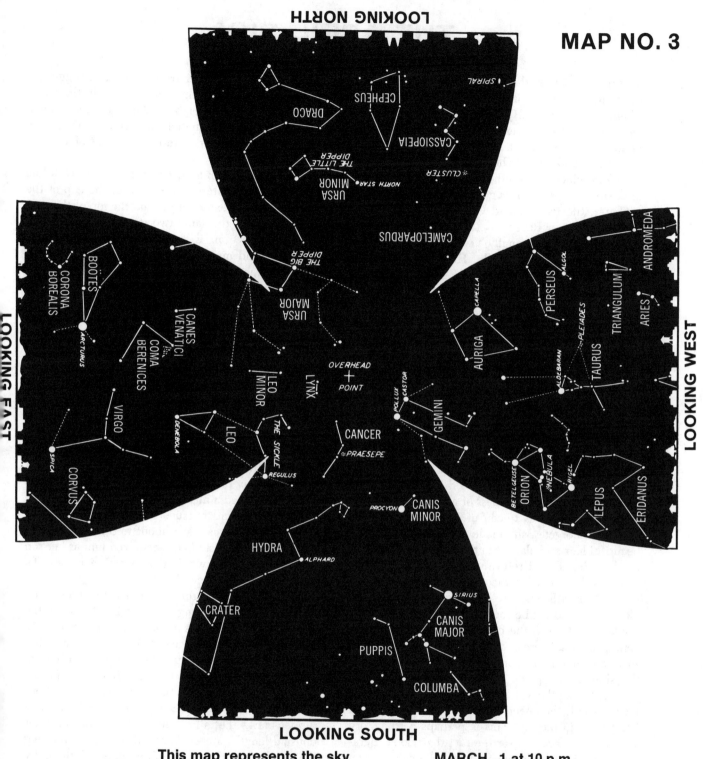

LOOKING SOUTH

**This map represents the sky
at the following standard times**

MARCH 1 at 10 p.m.
MARCH 16 at 9 p.m.
APRIL 1 at 8 p.m.

Arcturus and Aldebaran are about 8000°, Sirius and Rigel between 10,000° and 25,000°, and Spica is close to 35,000°. These are surface temperatures; in their centers, the temperatures of the stars are many millions of degrees. In terrestrial laboratories, man can surpass the temperature of the sun's surface; in a hydrogen bomb explosion, the temperature of the centers of stars is equaled.

THE APRIL SKY

STRETCHING ACROSS the southern sky from high in the southwest to low in the southeast is the longest of all constellations: Hydra, the Water Serpent. Its head lies below Cancer, and its body, with some bends in it, is easily traced by a line of faint stars below Leo and Virgo. The back of Hydra has been a favorite place to put other constellations. An Owl once stood there, but it is gone. Two groups remain — Crater, the two-handled Grecian Cup, and Corvus, the Crow or Raven.

The raven was originally a white bird and was a scout and gossip for Apollo. He would send the raven to spy on those whose affections he enjoyed, to make sure that they were faithful to him in his absence. On one occasion, the tale brought back by the bird was not a pleasing one, and Apollo flew into a rage, cursing the bird and condemning it forevermore to be black instead of white. And since that day ravens have been black. Relenting somewhat, Apollo must have ordered the bird to be placed in the sky, for we find it perched on the back of Hydra, where the stars which mark it form what seamen call "the Mainsail."

The constellation in which the north pole of the Milky Way is found is now high in the east. Coma Berenices is at once a tribute to the loving sacrifice of a queen and the cleverness of a court astronomer. When Berenice was queen of Egypt and wife of Ptolemy Euergetes, she made a vow to cut off her beautiful hair and place it in the temple of Venus, if her husband should return safely from his military ventures. She kept her pledge, but shortly after she made the sacrifice the tresses disappeared from the altar and could not be found. It was only the quick thinking of Conon, the court astronomer, that prevented a great uproar over the loss. He took the royal couple out of doors one night, and pointed to the shimmering patch of stars between Leo and Boötes, telling them that the gods had been so well pleased with the queen's gift that they had placed her tresses in the sky. This is perhaps a true story and this may account for the fact that Coma Berenices, the Hair of Berenice, was not generally accepted as a constellation when it was first suggested.

We can always say, of course, that in every myth there is at least a small grain of truth, if we can but find it. There are several very real persons represented in the heavens, however, one of them being a Stuart king, Charles II; the court physician said that a certain star had shone with especial brightness on the eve of the return of the king to London on May 29, 1660. Whether true or not, this suggestion led to the naming of the star Cor Caroli, the Heart of Charles. It is the brightest star in the inconspicuous group Canes Venatici, the Hunting Dogs, between Coma Berenices and the handle of the Big Dipper.

To find the North Star, remember, we draw a line through the two stars in the front of the cup of the Big Dipper and extend it beyond the top of the cup. A line through the same two stars, extended in the opposite direction, will encounter Leo, the Lion. The head and mane of the lion are marked by the Sickle, which has a bright star, Regulus, at the end of the handle. The star Denebola is in the tuft at the end of Leo's tail, and it is a good example of the corruption of an original Arabic expression, *al-dhanab al-asad*, the "lion's tail." With Arcturus and Spica, Denebola forms an equilateral triangle.

The Big Dipper in England is called the Plough or, sometimes, Charles's Wain. This latter is a corruption, however; it does not refer to a King Charles and his wain, or wagon; instead, it was originally the Peasant's, or Churl's, Wain. It is the conspicuous portion of Ursa Major, the Greater Bear. Callisto was a beautiful girl who aroused the jealousy of Juno and was punished by being transformed into a bear. She lived a miserable existence, afraid of the other animals and pursued by hunters. One day a young hunter, Arcas, was about to transfix her with his spear, for Callisto had recognized him as her son and had rushed toward him to embrace him. To prevent the matricide, Jupiter changed Arcas too into a bear; then, to protect them from further harm, he put them both among the stars. But Juno had the last word, for she went to Oceanus and Tethys, the rulers of the seas, and asked that Callisto and her son be forbidden to enter the waters. As a consequence, Ursa Major and Ursa Minor are in the northern heavens, where they circle endlessly about the North Star and never set. The handle of each dipper marks the tail of one of the bears, and their unusual lengths are explained this way: in hoisting the bears into the sky, Jupiter grasped them by their tails, and because of the great distance and the weights of the bears the tails stretched!

Intimately connected with the bears, Boötes was originally known as Arctophylax and is sometimes called the "Bear Driver." He holds in leash the Hunting Dogs, and as the rotation of the earth carries Ursa Major around the North Star Boötes seems to be pursuing the bear. When Carlyle wrote, "Over-

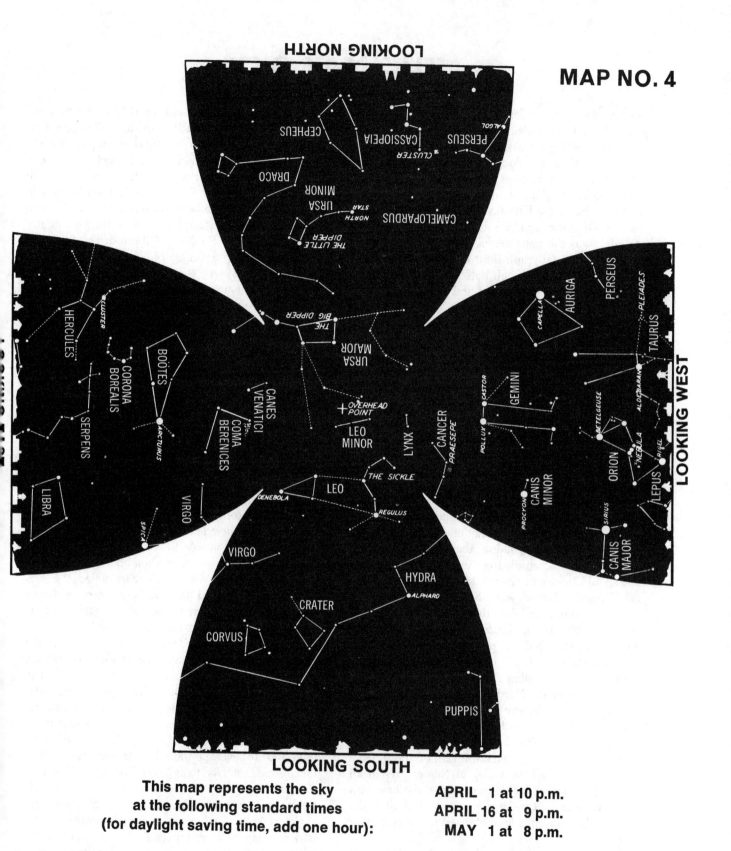

**This map represents the sky
at the following standard times
(for daylight saving time, add one hour):**

APRIL   1 at 10 p.m.
APRIL 16 at 9 p.m.
MAY   1 at 8 p.m.

head, as always, the Great Bear is turning so quiet around Boötes," he was contributing to the confusion which sometimes identifies Boötes as Arcas, who became Ursa Minor. The "Arcturus" in Job 38:32 of the King James Version should be Ursa Major, as it is in the Revised Version; and the "Orion" in Amos 5:8 should be Boötes. This constellation has been confused even with Erichthonius, the inventor of the chariot, but certainly Auriga must be the memorial to that benefactor of mankind.

THE MAY SKY

Low in the west the last stars of the winter are with us early in the evening, while at the same time all the east is filling up with the stars of summer. Now the Milky Way lies along the horizon, all the way around, and can be seen only if the observer is far from the lights of any of man's settlements.

Somewhat to the left of Arcturus is a lovely circlet of stars marking Ariadne's crown, Corona Borealis. Ariadne was the daughter of Minos, king of Crete, who kept in a labyrinth built by Daedalus a dread monster, half man and half bull, known as the Minotaur. Each year seven youths and seven maidens of Athens were exacted as a tribute by Minos, and were thrown into the labyrinth to be devoured by the Minotaur. The victims were always chosen by lot, but at last Theseus, the son of King Aegeus of Athens, volunteered to go as one of them, in the hope of being able to slay the monster. When the black-sailed ship arrived with its cargo of sacrifices, Ariadne saw Theseus and fell in love with him. She gave him a sword; more important, however, she gave him a skein of thread, which he was to unwind behind him as he entered the labyrinth, then follow again in order to escape. Of course, Theseus slew the Minotaur and escaped with his companions, taking Ariadne with him. They stopped at the island of Naxos, where Theseus abandoned Ariadne as she slept. On waking, Ariadne abandoned herself to grief, but Venus consoled her with the promise that she should have an immortal lover, in place of mortal Theseus. Naxos was the favorite retreat of Bacchus, who found her there and made her his wife. He gave her a golden crown, set with gems, and when she died he placed the crown in the heavens, where we find it in these evenings, between Boötes and the kneeling Hercules.

Even as an infant, Hercules strangled two serpents sent by Juno to destroy him. As a further mark of Juno's hostility, he was bound to serve Eurystheus, who gave him twelve dangerous labors to perform. The first was to bring back the skin of the terrible lion which roamed the valley of Nemea. When all his weapons failed, Hercules strangled the lion with his bare hands. Incidentally, it is probably the Nemean lion which is immortalized in the constellation Leo.

The second labor of Hercules was the slaughter of the many-headed Hydra, which guarded the well of Amymone. Another was the cleansing of the stables of King Augeas of Elis; in them, three thousand oxen had been housed, yet for thirty years the stables had not been cleansed, until Hercules diverted the rivers Alpheus and Peneus, to make them flow through the stables and purify them in one day. On another occasion he had to bring back to Admeta, the daughter of Eurystheus, the jeweled belt of Hippolyta, queen of the Amazons. Then he had to bring back the three-bodied oxen of Geryon, and on this adventure he gave his name to the two mountains forming the Straits of Gibraltar: the Pillars of Hercules. The purloining of the golden apples of the Hesperides was the occasion for Hercules to hold up the heavens, giving poor tired Atlas a brief rest. The legend of Hercules is very old; we have seen how even before the days of Greek story there were tales of a prodigiously strong giant who struggled with a dragon. In the sky, Hercules is shown on one knee, with his other foot on the head of a dragon.

By referring to Map No. 6, we see the diamond-shaped head of Draco, the Dragon, whose body turns first toward Cepheus, then doubles back to wind between the two dippers. The tip of the tail lies almost on the line from the Pointers to the North Star. The stars are all only moderately bright, but the line is easily traced.

In the western edge of Hercules (on our maps the area is labeled "cluster") there is an object which is faintly visible to the naked eye and is a most magnificent spectacle when photographed with a great telescope. We have earlier noted the Pleiades, Praesepe, and the patch of stars in Coma Berenices; these are of the so-called "galactic" type of cluster, because most of the 300-odd known objects of this class are found in or very near the stream of the Milky Way. We know more than a hundred so-called "globular" clusters, in addition to the open or galactic objects; the object in Hercules is one of the fine examples. Whereas in a galactic cluster we find from perhaps twenty to a few thousand stars, a globular cluster will contain many thousands to several hundreds of thousands. A galactic cluster is referred to as "open," or "loose," because the stars in it seem not to be arranged in any particular way. But in a globular cluster the stars form a nearly spheroidal mass in which practically no dark sky can be seen, especially toward the middle, where the stars are apparently so close together as to be separable only with great telescopes, and even then there is a haze or fog from the combined light of the thousands of stars too faint to be individually distinguished, even with long exposures.

LOOKING NORTH

LOOKING SOUTH

LOOKING WEST

**This map represents the sky
at the following standard times
(for daylight saving time, add one hour):**

**MAY 1 at 10 p.m.
MAY 16 at 9 p.m.
JUNE 1 at 8 p.m.**

These globular clusters seem to be so distributed in our galaxy as to outline its central bulge. Because we are not in the center, but more than halfway from the center toward one edge, we see most of these clusters in one half of the sky. The center of this half is now rising in the southeast; it is just about at the point where the modern boundaries of the constellations Ophiuchus, Sagittarius and Scorpius meet, above the sting of the Scorpion, where rich star clouds are found.

27

THE JUNE SKY

BRILLIANT ARCTURUS now stands high in the south, while Vega, yet brighter, pulls the eyes to the east. Low in the south the Scorpion stands, with its heart marked by red Antares ("anti-Ares" — the rival of Ares, or Mars).

The zodiacal constellations now above the horizon wholly or in part are, from west to east, Gemini, Cancer, Leo, Virgo, Libra, Scorpius and Sagittarius. On June 22, when summer begins, the sun stands in the direction of the feet of Gemini, now below the western horizon. When autumn begins, September 23, the sun is in the direction of a point in Virgo almost halfway from Spica toward Regulus in Leo. At the beginning of winter, December 22, the sun stands in Sagittarius above the tip of the tail of Scorpius.

Libra, the Balance, is the only inanimate object in the zodiac. We have seen earlier that these stars were once known as Chelae, the Claws (of Scorpius); the two brightest stars of Libra yet bear the names Zubenelgenubi, the Southern Claw, and Zubeneschamali, the Northern Claw!

Libra may be linked with the neighboring constellation Virgo, for one legend has it that she represents Astraea, the goddess of justice, the last to leave the earth after the golden and silver ages, as man grew more violent and brutal. Another identification for Virgo is with Ceres, the goddess of the harvest, and the star Spica represents a stalk of grain which she holds in her hand.

In the west the stars of early spring are leaving us; Hydra's head is setting while the end of the tail is due south! In the north, the W of Cassiopeia is now right side up, while high in the northwest the Big Dipper starts downward. The large vacant area between Ursa Major and Perseus is occupied by the constellation Camelopardus, the Giraffe. It now stands upright, but it will hardly be possible to pick out the figure of the creature, with so few faint stars in this area.

In Cepheus, one corner of the figure is marked by a small triangle; the star at the sharpest corner of this triangle is the prototype of an interesting and important class of variable stars known as the Cepheids. These stars, unlike the eclipsing variables, are intrinsically changeable. A Cepheid is a single star which is alternately expanding and contracting, in a definite pattern or rhythm, repeated exactly each time; we call such action "pulsation." When a Cepheid is contracting, and is about at its average (that is, halfway between maximum and minimum)

diameter, it is at its faintest; when it is expanding and is at its average diameter, it is at its brightest. A fair average amount of increase of light for a Cepheid is about 100 per cent; that is, the star is twice as bright at maximum as it is at minimum. This range is exceeded by many Cepheids, however.

Because of the regularity of the variations of a Cepheid and the particular pattern of its changes, such a star can be recognized, no matter where it may be located. Some of them are near enough to us to have their distances measured, and as soon as we know a star's distance and its apparent brightness we can determine its real brightness as compared, say, with the sun. When this is done for Cepheid variables, it is found that the average intrinsic brightness of one with a period of one day or less is about fifty times the sun's luminosity; one with a period of two days is about six hundred times the sun's brightness; a period of fifty days indicates a luminosity of fifteen thousand times that of the sun. Of course, those with intermediate periods have intermediate luminosities.

This "period-luminosity relation" for Cepheids is one of the most powerful tools we possess for determining great distances. Suppose we find in a globular star cluster or remote galaxy some Cepheid variables. We shall be able to recognize them as has been explained, and after some time we shall have accurate determinations of their periods and apparent brightnesses. Then, using the periods with the period-luminosity relation, the intrinsic brightnesses can be obtained; when these are compared with the apparent brightnesses, the distances can be derived. By this means, distances of thousands or even millions of light-years have been determined with comparatively small uncertainties.

Another type of variable star is the so-called nova, or "new star." Not really new, such a star represents an extreme instance of cataclysmic expansion: the star literally swells up and bursts! We must think of stars as being in rather delicate balance: radiation from a star's interior must struggle against the opacity of the overlaying layers in order to escape. If the rate of generation of energy is too great for the outer layers of the star to permit to pass, the star will swell up, at first slowly, and perhaps this will let the excess energy escape. The star then may collapse, then expand again, and thus become something like a Cepheid.

But when the release of energy is too violent, the star will explode, and for a brief while will shine

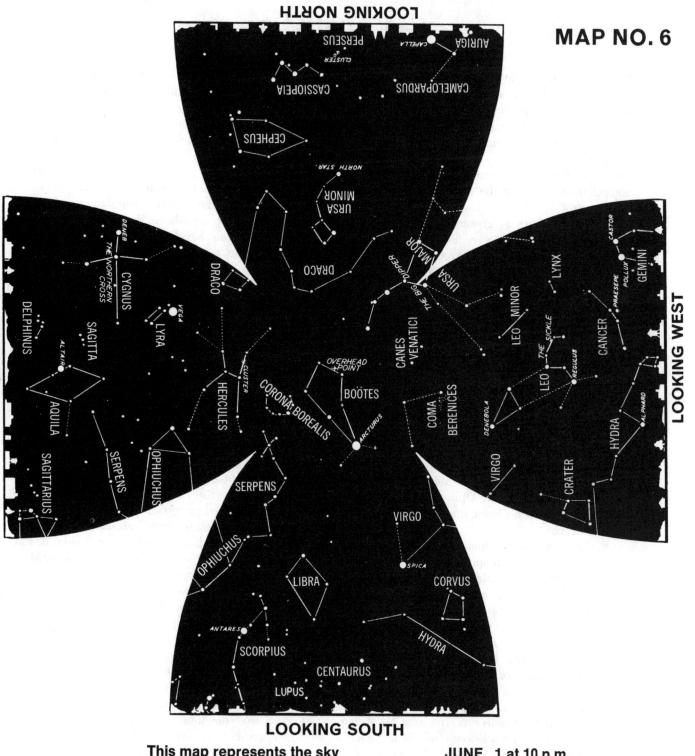

LOOKING NORTH

LOOKING SOUTH

This map represents the sky
at the following standard times
(for daylight saving time, add one hour):

JUNE 1 at 10 p.m.
JUNE 16 at 9 p.m.
JULY 1 at 8 p.m.

with a luster many thousand times its original brightness. Such a star appeared closely northwest of Vega at the end of January 1963; by April it had faded to invisibility to the naked eye. At maximum it was about half as bright as the North Star. Several fainter novae are discovered each year. The amateur who takes a careful look at the Milky Way each clear night may be the fortunate discoverer of a nova.

THE JULY SKY

FOR THOSE who live in the northern latitudes of the United States, Scorpius is now seen to best advantage. Antares is in the body of the arachnid, the stars in the string up to the right are the shortened claws, and the tail of the Scorpion curves down toward the horizon, then back up, to end in two stars marking the sting. For observers in southern latitudes, this constellation is one of the most magnificent in all the heavens.

An amusing turn is given one of the old star groups when we make a very fine teapot of some of the stars of Sagittarius, the Archer, who is traditionally pictured as a centaur, with the upper part of the body of a man affixed to the body of a horse. Noted for cunning and wisdom, and learned in the arts, the centaurs were the mentors of the gods and their half-mortal offspring.

Spreading across the sky from south to north is an array of constellations constituting what must be an enormous allegory. Hercules is crushing the life out of Draco, above and in the north; in the south, another great figure of a man holds a serpent in his hands and treads on the Scorpion. Certainly here we have some attempt to portray the triumph of good over evil; Hercules and Ophiuchus were good men and Draco and Scorpius are loathsome creatures. The brightest star in Ophiuchus, at the top of the triangle, marks his head; the head of Hercules is the star nearby.

Of the goodness of Ophiuchus we have ample evidence. He was Aesculapius, son of Apollo, and a pupil of Chiron, the centaur represented in Sagittarius. He became a great physician, so successful in saving lives that Jupiter had to slay him, to put an end to the grumbling of Pluto, the ruler of Hades, because the flow of souls to the underworld was dwindling. But after his death he was received among the company of the gods.

The snake as an emblem of health was sacred to Aesculapius, because of its apparent renewal of life by periodic shedding of its skin. In classic art, Aesculapius is pictured with a heavy staff, about which a serpent is twined. It was his official badge, and so it is the emblem of medicine even today. But in some regrettable way this symbol is often confused with the caduceus of Hermes or Mercury, a thin wand with a pair of wings at the top, and two thin snakes draped gracefully and symmetrically about it. This latter emblem has nothing to do with medicine; it is properly the symbol of messengers, not medical men.

The serpent that Ophiuchus grasps in his hands is Serpens, whose head is an X of stars south of Corona Borealis and whose tail is near Aquila, the Eagle. Because this long constellation is interrupted by Ophiuchus, it is sometimes (as in the excellent Becvar atlas mentioned earlier) divided into Serpens Caput (head) and Serpens Cauda (tail).

When Hercules at last died and went to dwell with the gods, Juno's wrath toward him was softened, and she gave him her own daughter Hebe in marriage. Hebe had been the cup-bearer to the gods, and upon her retirement Jupiter disguised himself as an eagle and swooped down to earth, seizing the Trojan youth Ganymede and carrying him to Olympus to be the new cup-bearer. In the constellation Aquila, this eagle of Jupiter is immortalized.

To the left of Aquila, in the eastern sky, the Northern Cross lies on its side. It is the conspicuous portion of the constellation Cygnus, the Swan. The story of Cygnus is the story of Phaëthon, the son of the sun-god Helios and the nymph Clymene. He didn't know who his father was until one day his mother told him, to console him for the taunts of his companions. He went to seek Helios, and when he found him he was welcomed and promised any boon he might ask. Quickly he demanded to be permitted to drive the chariot of the sun for one journey. Helios tried to dissuade him, but Phaëthon insisted, and at last he started out along the dangerous course no one but Helios had ever followed. The way at first was not so difficult, but soon the spirited steeds realized an inexperienced hand held the reins, and they ran wild. Plunging too close to the earth in some parts of the path, there were great fires started, and great parched deserts were formed; the world was threatened with destruction until Jupiter launched a thunderbolt which struck Phaëthon and hurled him from the chariot. He fell into the river Eridanus, and his friend Cycnus tried to find his body. Cycnus dived into the river and swam to and fro, sometimes with his head beneath the water, until at last the gods took pity on him and changed him into a swan. Even today, we see swans frequently putting their heads below the surface of the water, perhaps still looking for Phaëthon. Later, Cycnus was transported to the heavens, where we find him as the constellation Cygnus, the Swan.

Many feel that the legends are nature myths in which little-understood phenomena have been interpreted as the doings of the gods. The legend of Phaëthon is supposed to "explain" a period of

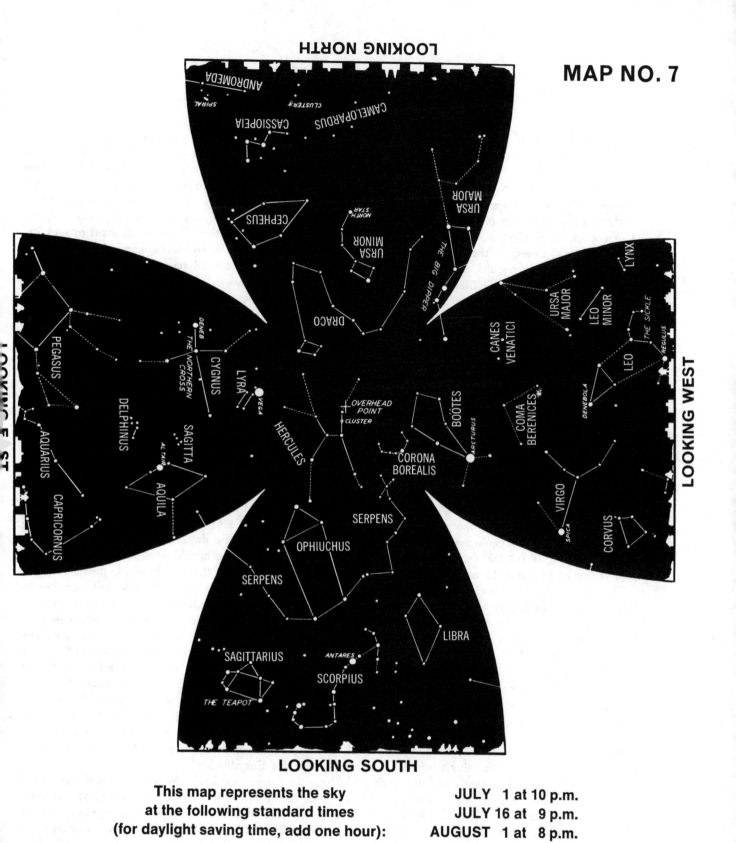

MAP NO. 7

LOOKING NORTH

LOOKING SOUTH

LOOKING WEST

This map represents the sky at the following standard times (for daylight saving time, add one hour):

JULY 1 at 10 p.m.
JULY 16 at 9 p.m.
AUGUST 1 at 8 p.m.

drought. There are others who believe the myths are intended as parables, to teach moral values. In this view, Phaëthon would be a youth whose overconfidence should have yielded to better judgment; when it did not, he caused great hardship to others before he was punished by Jupiter.

The star Deneb marks the tail of the Swan, while the star Albireo, at the foot of the Northern Cross, marks the beak of the bird as it flies along the Milky Way.

THE AUGUST SKY

AUGUST IS the month when the brightest summer stars shine directly overhead in the night sky. Brightest of all, Vega is almost in the zenith at the hours for which Map No. 8 is drawn. The bright summer Milky Way arches high, from Perseus in the northeast through Cassiopeia, Cepheus, Lacerta (not shown on the map), Cygnus, Aquila, Sagittarius, and Scorpius. But at Deneb, in the head of the Northern Cross, the Milky Way splits, and a fainter branch runs through the head of Ophiuchus.

When we gaze toward Sagittarius and the sting of Scorpius, we are looking in the direction of the center of our galaxy. There the greatest concentration of stars can be seen, and undoubtedly equally large numbers are hidden by obscuring clouds of interstellar gases and dust. In fact, the bifid character of the Milky Way between Cygnus and Scorpius is due to dark nebulae and not to an absence of stars.

The size and shape of the stellar aggregation we know as our galaxy have been described earlier. But this is not a static system; there is motion in it. The galaxy is rotating in its own plane, in such a way that the sun and its attendant family of planets make a complete swing around the distant center once each 225,000,000 years. This may seem to be a very long period of time, but geologists put the age of the earth at about 5,000,000,000 years, so we have made about twenty trips since the earth was born!

The sun and its neighboring stars all partake of this rotation of the galaxy at about the same speed, approximately 175 miles per second. Meanwhile, however, each star is moving somewhat with respect to the others, as each gnat in a swarm moves with respect to the others while the whole swarm moves as a unit. By this motion the sun is traveling at a rate of about 12 miles per second toward a point between Vega and some of the faint nearby stars of Hercules.

Vega is the brightest star in Lyra, the Lyre of Orpheus, upon which that singer of sweet song played such magic music that even the birds and beasts yielded to its charm, and waterfalls ceased splashing while the music played. Orpheus loved and was beloved by Eurydice, and they were wed. One day, while playing with her companions in a flowery meadow, Eurydice stepped on a small serpent and was bitten in the foot and died. Her soul went to the underworld, and Orpheus wandered through the earth disconsolate, playing such sad music that all who heard it were moved to tears. At

length one day as he was pouring out his grief in song by the bank of a little brook, he heard a voice murmuring to him; it was the voice of the stream itself, and it told him that for a part of its course it had flowed through the underworld, and there it had seen Eurydice, and she too had been weeping. Emboldened by this, Orpheus resolved to seek entrance to the underworld. He charmed his way along the dank and dismal passages, past Cerberus, the horrible three-headed dog, and at last into the audience chamber of dread Pluto himself. There he sang his love for Eurydice in such strains that even Pluto's heart was melted, and he gave permission for Orpheus to lead Eurydice back to light and life again, but only on one condition: Orpheus should not look back at her until once more they had gained the outer world and were in the sunlight again. To this Orpheus eagerly assented, and he almost kept his promise; but only a step before the outer portal he turned to be sure she was still following, and she was instantly snatched from him and taken back to death, this time forever.

Once more Orpheus wandered about the world unhappily, and in Thrace there were maidens who tried to woo him from his grief, but he spurned them. At last one day, excited by the rites of Bacchus, the maidens hurled javelins and stones at him, but so great was the power of his music that the missiles stopped in mid-flight and fell harmless to the ground. But finally the maidens screamed so loudly that the music was drowned out, and then the stones took effect, and Orpheus was slain. His body was torn to pieces and thrown into the river Hebrus, but the Muses gathered the fragments and buried them at Libethra; and even today, it is said, the nightingale sings most sweetly in Thrace, over the grave of Orpheus. His soul went to join Eurydice in Tartarus, and the gods put the lyre into the sky, as a symbol of his sweet music and his great love for Eurydice.

In August we have almost completed the circuit of the heavens, for the last two zodiacal constellations are above the horizon. The sprawling V of Capricornus is better seen on Map No. 9, where it is almost due south. This figure is a "Sea Goat," whatever that might be. The front legs and head of the goat are affixed to the tail of a fish. Some believe this to be an outright error, for no such figure appears in ancient mythology. It may represent Pan, who once jumped in a river and assumed the form

MAP NO. 8

LOOKING NORTH

LOOKING SOUTH

LOOKING WEST

This map represents the sky
at the following standard times
(for daylight saving time, add one hour):

AUGUST 1 at 10 p.m.
AUGUST 16 at 9 p.m.
SEPTEMBER 1 at 8 p.m.

of a fish, to escape the consequences of one of the pranks in which he was always engaged.

Between Capricornus and Pisces lies Aquarius, the Water Bearer. He is usually represented as an old man, pouring water out of a jar and into the mouth of Piscis Austrinus, the Southern Fish.

THE SEPTEMBER SKY

Now THE Northern Cross is in the zenith, Arcturus is setting north of west, and the Pleiades are rising north of east. There will be many who wonder about the flashing star rising in the northeast; when Capella rises it sparkles and burns, because its light is disturbed by the tremulous air close to the horizon. It is true for any star, of course, but for Capella the effect is exaggerated, because that star is so far north that it takes a long time for it to rise high enough to shake itself clear of the dense and dirty air near the horizon. In December the same thing will be true in reverse for Vega as it sets far north of west.

Here and there we find flat contradictions among the classic myths. For example, one story of the death of Orion is that he was shot inadvertently by an arrow of Diana. Apollo, her brother, suspected the two were in love, and one day as Orion was wading in the sea with only his head above water Apollo pointed out the dark object and challenged Diana to send an arrow through it. Her aim was true; the dead body of Orion washed ashore.

Still another story of the death of Orion relates that he boasted that he was able to overcome any man or beast, whereupon a scorpion came out of the ground and stung him, with fatal effect. This ancient story is cited to justify the placement of the constellations Orion and Scorpius; as one rises, the other sets. The Scorpion now is almost down, in the southwest, while in a little time Orion will rise, due east.

If Map No. 9 is held so the words "The Northern Cross" are right side up, Pegasus will be seen to form a very fine horse. At least, the head, shoulder and forelegs are shown; a triangle marks the head, the Great Square marks the shoulder, and two lines of faint stars mark the legs, with the hoofs pounding on the Milky Way. One star of the Great Square belongs not to Pegasus, but to Andromeda, according to the modern constellation boundaries.

The origin of Pegasus, the Winged Horse, is connected with the adventures of Perseus to be related in connection with Map No. 12. From the severed neck of Medusa drops of blood fell to the earth, and from them sprang the wondrous horse. He was caught and tamed by the goddess Minerva, and was presented by her to the Muses, who provided inspiration for poets and other artists.

In Lycia there raged a fearful monster known as the Chimera. Its forequarters were compounded of a goat and a lion, and its hindquarters were those of a dragon. Breathing fire, it wrought much woe, and King Iobates sought a hero who would destroy it. There came to him the brave youth Bellerophon, who brought with him a letter of recommendation which had a postscript to the effect that Proteus, the son-in-law of Iobates, would consider it a great favor if Bellerophon should be put to death! To get rid of the Chimera, and perhaps at the same time to get rid of Bellerophon, Iobates set him the task, and the warrior accepted the challenge. First he consulted the soothsayer Polyidus, who advised him to use Pegasus and suggested that he spend a night in the temple of Minerva. He did so, and Minerva gave him a golden bridle and showed him where to find the winged horse. When he found Pegasus, the horse came to him willingly. Bellerophon mounted Pegasus, soon found the Chimera, and destroyed the monster.

Iobates, still anxious to get rid of Bellerophon, set him other dangerous tasks, but with the help of Pegasus the young warrior was successful in all of his ventures, except his last one. He grew vain and arrogant, and mounted on Pegasus he even tried to ride up to heaven, but Jupiter sent a gadfly to sting Pegasus, and Bellerophon was thrown to the ground and became lame and blind. He wandered in loneliness, avoiding all his former companions, and at last died miserably.

Below Pegasus, in the south and east, we find the "watery" constellations. The strange Sea Goat has already been mentioned, and below it, so close to the southern horizon that it can not be seen well from the northern United States, is the wading bird Grus, the Crane. Aquarius, the Old Man with the Water Pot, pours a stream of water into the mouth of Piscis Austrinus, the Southern Fish. In this latter group, the very bright star Fomalhaut is located; as a matter of fact, there is hardly any more to the constellation than that one bright star; it is so isolated that it always attracts attention. Below and east of Pegasus is the constellation of Pisces, the Fishes. One fish is due south of the Great Square, the other is due east; their tails are tied together with a long ribbon which forms a V-shaped group of faint stars well shown on the map. South of Pisces is the Sea Monster, Cetus. All these groups have some connection with the water, and they lie close together in the sky because of this common connection, in all likelihood.

Returning to the southern sky, we see high up the

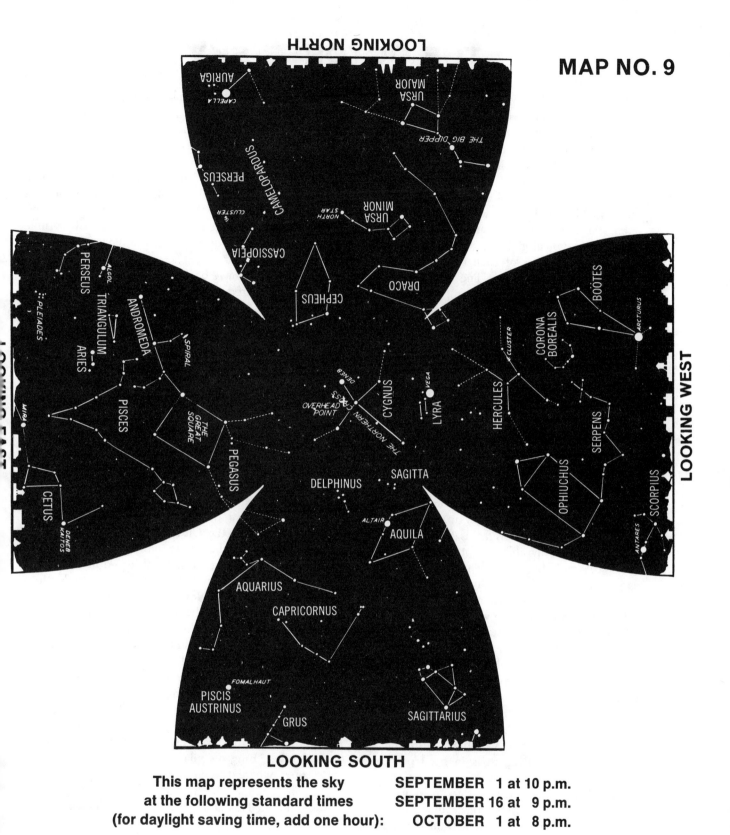

LOOKING NORTH

LOOKING SOUTH

This map represents the sky   SEPTEMBER   1 at 10 p.m.
at the following standard times   SEPTEMBER 16 at   9 p.m.
(for daylight saving time, add one hour):   OCTOBER   1 at   8 p.m.

little constellation of Delphinus, the Dolphin. According to one legend, the dolphin persuaded Amphitrite to yield to the wooing of Neptune and was placed in the sky by the grateful ruler of the seas. Another myth makes the dolphin the rescuer of the poet-musician Arion.

THE OCTOBER SKY

AGAIN WE HAVE the sky in a transition stage. In the west the fine summer stars have almost gone, while the east is beginning to fill up with the stars we shall be seeing all winter. There are a great many interesting objects in this sky, however, and we can well spend our time now in reviewing them.

The two typical star clusters are now simultaneously visible; in the east we see the Pleiades, an open star cluster, while in the west, in Hercules, there is the typical globular star cluster. There are other representatives of each type in the October heavens, however; another fine open cluster is in Scutum (not shown on the map), between Aquila and Sagittarius, and another fine globular cluster is in Pegasus, just off the tip of the horse's nose. Many of the approximately one hundred globular star clusters are faint, and most of them are concentrated strongly toward the center of the galaxy; in one photograph made at the Southern Station of the Harvard College Observatory, with the sting of the Scorpion approximately in the center of the field, there are thirty-four globular clusters shown, a third of all that are known!

The interesting variable star Delta Cephei, in the triangle at one corner of the figure of Cepheus, is now almost overhead; another interesting intrinsic variable, Mira Ceti, is also well up. At its maximum, Mira is sometimes as bright as the North Star; at minimum, it is invisible except in telescopes. This star, too, is pulsating, and it changes in temperature as well as in size. Incidentally, it is one of the largest known stars; its diameter is of the order of 300,000,000 miles, a figure which we might compare with the diameter of the sun, which is 864,000 miles. Mira is the type-star for the so-called long-period variables; its period is about 330 days, although it may be a little more or a little less, for any particular cycle. Long-period variables are considered to be regular, but their precision of repetition is not as high as that of the Cepheids.

The typical eclipsing variable is also above the horizon. Algol, in Perseus, is in good position for observation.

The places of several new stars, or novae, are in the sky at this time. Most spectacular was the one seen by Tycho Brahe, the great Danish astronomer, on November 11, 1572; others had seen it a few days earlier, but Tycho studied it carefully. For example, he determined by observation that this star was definitely not a phenomenon of the earth's atmosphere but was instead a prodigy among the fixed stars. It appeared in Cassiopeia and for a while shone more brilliantly than any other object in the heavens, excepting the sun and the moon; it was easily visible to the casual eye in broad daylight.

In Perseus, in 1901, there appeared a nova which at its maximum rivaled the most brilliant stars, and one only somewhat fainter appeared in Aquila in 1918. In Hercules in 1934, and in Lacerta (not shown on the map), between Cepheus and Cygnus, in 1936, there were novae which were easily visible to the unaided eye. The one near Vega in 1963 was described earlier.

What must have been one of the most extravagant novae appeared about nine hundred years ago and seems to have been recorded only by the Chinese; at about this time, Europe was intellectually dormant. Today we see with telescopes an interesting nebula near the tip of the lower horn of Taurus; it is shown best in photographs and was named by Lord Rosse the "Crab" nebula. By means of photographs made at long intervals of time, it has been found that the nebula is slowly expanding, and if we carry back with the average rate of expansion to the moment when the nebula was all at one point, it is found to have taken almost nine centuries for the nebula to attain its present dimensions. Now, in the year 1054 A.D. the Chinese observed a very brilliant nova in or very near this same spot, and we are firmly convinced that the Crab nebula represents the remains of that stellar explosion. We are confirmed in this belief by what has been observed in connection with certain novae in modern times. Both Nova Aquilae 1918 and Nova Persei 1901 threw off expanding gaseous shells which were later seen to be small nebulae, the difference between them and the Crab nebula being one of size and not of kind.

We find other nebulae which we have become convinced originated from novae explosions; because of their small, compact appearances we call them "planetary nebulae." Some of them take the forms of annuli, or rings, much like doughnuts in appearance; the best example of this type is the so-called "Ring" nebula in Lyra, visible only in telescopes, where it resembles a smoke ring.

The spiral in Andromeda is again well up, and should be looked for on moonless nights. In this neighboring galaxy there appeared in 1885 a nova which was easily visible in a small telescope. Because of its enormous distance as compared with the distance of even the most remote novae in our own galaxy, in order to be even so bright it must have had an amazingly high intrinsic luminosity, and this and other such objects we now call "supernovae."

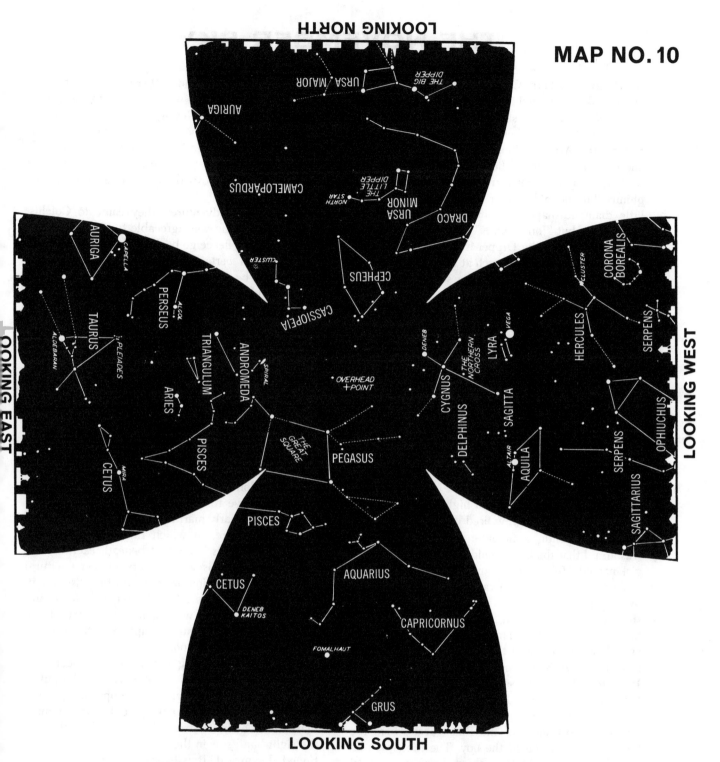

MAP NO. 10

This map represents the sky
at the following standard times
(for daylight saving time, add one hour):

OCTOBER 1 at 10 p.m.
OCTOBER 16 at 9 p.m.
NOVEMBER 1 at 8 p.m.

THE NOVEMBER SKY

Now THE Northern Cross stands upright in the western sky, and bright Vega and Altair are below it. Between the latter star and Albireo, the star at the foot of the cross, lies the small constellation of Sagitta, the Arrow. Whose arrow it is we do not know; if it came from the bow of Sagittarius, the Archer, it has gone far astray! Above Altair is Delphinus, but no old farmer or sailor thinks of the little diamond-shaped portion of this constellation as anything but "Job's Coffin."

In the north the Big Dipper is in its most unfavorable position, with Benetnasch at the end of the handle scraping the horizon. In the south, Fomalhaut in Piscis Austrinus, the Southern Fish, rides in almost solitary splendor, while higher and farther toward the east Cetus is now well displayed.

Driven in as a wedge between Andromeda and Pisces on the one side and Aries on the other is the small constellation of Triangulum, the Triangle. Why the ancient namers of the constellations should have become suddenly so literal is somewhat puzzling; perhaps the Greek letter delta, which in upper case looks like a triangle, had something to do with it. Very early maps and descriptions speak of this constellation as "Deltoton."

Aries is the Ram of the Golden Fleece which was the object of great search and high adventure. King Athamas in Thessaly grew tired of his wife Nephele and put her aside in favor of another. The queen was afraid that the king would harm her children, a girl named Helle and a boy named Phrixus, and wished to send them far away. Mercury assisted her, by giving her a ram with golden fleece, on which she set the two children, trusting the ram to carry them to a safe place. The ram leaped into the air, and all went well until they were crossing the narrow strait which divides Europe from Asia, when Helle fell off into the sea. Since that day the strait has been called the Hellespont.

The ram, carrying Phrixus, continued to Colchis, on the eastern shore of the Black Sea, where the king, Aeëtes, welcomed the boy. The ram was sacrificed to Jupiter, but the golden fleece was placed in a sacred grove, guarded by a watchful dragon.

In another part of Thessaly another king, Aeson, surrendered his throne to his brother Pelias, who was to rule as regent until the majority of Jason, who was Aeson's son. When Jason became a grown man and demanded his crown of Pelias, his uncle thought he might manage to get rid of his unwanted nephew by sending him on a quest to return the golden fleece to Thessaly, where it had originated. Jason fell in with the idea and gathered about himself a great company of noted warriors. Among them were Hercules, Orpheus, Nestor, Theseus, and many others who were destined to become known for their later exploits. The ship for the expedition was built by Argos, and was called the Argo; from its name the members of the expedition became known as the Argonauts.

After several adventures, they came to Colchis, and there the king was agreeable to the idea of letting the golden fleece go back to Thessaly, but Jason must perform certain dangerous tasks before the fleece could be obtained. We do not need to explain that he and his companions, by means of certain miraculous interventions and charms, succeeded in carrying the golden fleece back home, where the good ship Argo was dedicated to Neptune, who placed it in the sky. Most of this constellation lies below the horizon for most of us in the United States, but reference to Map No. 2 will reveal a portion of Puppis, the stern, or "poop," of the ship. Incidentally, in the sky the ship Argo has no bow, because very early in their adventure the Argonauts sailed between the Symplegades, or Clashing Islands, and there the bow was crushed. Whatever became of the golden fleece we do not know, but most scholars believe this story to be the legendary recital of a very early maritime expedition, perhaps of a piratical nature, in which rich spoils were obtained. Again, it is likely that beneath each mythological veil there is a true story of some natural occurrence or occasion, if we could but discover it. Some suggestions are rather far-fetched; for example, there are a few who believe the story of the Argonauts to be a pagan corruption of the story of Noah and the Great Flood.

Most of the interesting myths have now been told, but in December the "Royal Family" will be put in the spotlight. Hold the map on the opposite page in the proper position for "Looking South"; let an imaginary line be drawn from the North Star through the right-hand star in the W of Cassiopeia and continued downward. It will pass just inside the left-hand or eastern edge of the Great Square of Pegasus; if it is continued downward a distance equal to the length of this side of the Great Square, it will mark the spot in which the sun stands on March 21, the date of the beginning of spring. On this day the sun passes north of the celestial equator, after having spent six months on the south side of it. This point, called the vernal equinox, is in the constellation Pisces, the Fishes.

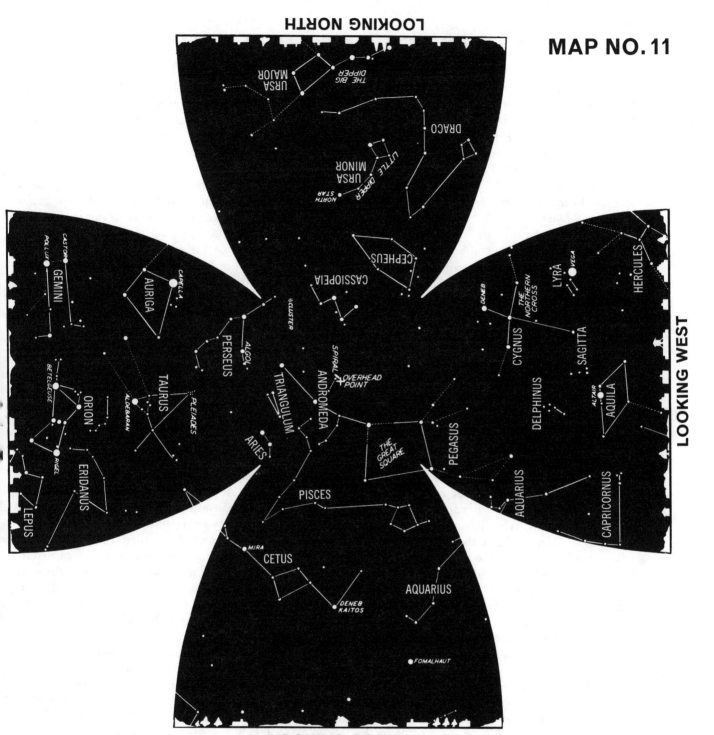

LOOKING NORTH

LOOKING WEST

LOOKING SOUTH

This map represents the sky
at the following standard times

NOVEMBER 1 at 10 p.m.
NOVEMBER 16 at 9 p.m.
DECEMBER 1 at 8 p.m.

THE DECEMBER SKY

AGAIN THE Northern Cross stands erect in the west, now almost on the horizon. A little way to the north, Vega is setting. In the south, the large comparatively vacant area between Eridanus and the bright solitary star Fomalhaut precedes the brilliant Orion region in the southeast. Perhaps it may be well to point out that the word Fomalhaut is not French and should not be pronounced *fo-mal-hoe;* instead, it is *fo-mal-hawt.*

The triangle formed by Betelgeuse in Orion, Procyon in Canis Minor, and Sirius in Canis Major now stands on its base, in the east, while Capella in Auriga is high up, approaching the zenith. In the north, the Big Dipper is almost on the end of its handle, beginning another round; the Little Dipper, with handle bent backward, is hanging down from the North Star.

Above the Little Dipper, looking like an M as we gaze north, is Cassiopeia, a queen of Ethiopia who was beautiful and was too vain about it. She would sit by the hour, combing her lovely hair and gazing at herself in a mirror. One day she boasted that even the sea nymphs were not as beautiful as she, and the word got to these maidens, who were provoked by it to such a degree that they persuaded the ruler of the waters to send a horrible sea monster to ravage the coast of the country in order to punish Cassiopeia for her vanity. When the conditions had grown quite intolerable, Cepheus, the somewhat subdued king, went to an oracle, who told him that he could rid his country of the horrid pest only by exposing his daughter Andromeda to be devoured by the monster. Tearfully he ordered Andromeda to be chained to the rocks, and there we must leave her, in imminent danger of being killed by the scourge of the country, while we pick up another thread of the story.

Perseus was the son of Jupiter and Danaë. Acrisius, his grandfather, was alarmed by the prediction of an oracle that his daughter's son would be the instrument of his death, and he had them both locked in a chest and thrown into the sea. They were rescued by a fisherman, who took them to Polydectes, a king in Seriphus, who received them warmly. When Perseus had come to manhood, he was sent to slay Medusa, a horrible creature which had once been a beautiful maiden who had dared to compare features with those of the goddess Minerva. The jealous goddess had taken Medusa's charms

from her, changing her lovely teeth to brutish fangs and her lustrous hair to writhing serpents. So frightful was her new appearance that all who gazed at her countenance were changed to stone. Her cavern was filled with stone figures which had once been living things.

When Perseus set out to destroy this wretched creature, he was aided by Minerva, who lent him a brightly burnished shield. Perseus must not look directly at Medusa; he must see only her reflection in the mirrorlike surface of the shield, lest he be turned to stone. A second loan of great importance came from the god Mercury; it was a pair of winged sandals, to speed Perseus on his journey.

Thus equipped, he set out to find Medusa. He had to ask the Graeae, three ugly sisters who had only one eye among them which they passed from one to another. By intercepting the eye in transit, Perseus had them at his mercy and forced them to tell him the way. He found Medusa sleeping, and with one quick stroke of his weapon, while gazing at her reflection in the shield of Minerva, he severed her head from her body and put an end to the poor creature as hateful to herself as to the rest of the world.

On his way home, he stopped for food and rest in the realm of Atlas. There he was not very warmly welcomed, because Atlas too had been warned that a son of Jupiter would cause his loss of the Golden Apples which hung in his garden. Perseus lifted the head of Medusa from the bag in which he carried it, and let Atlas gaze on it; the monarch was changed into stone, and the heavens rested on him, as they do to this day. You may see this giant king in the Atlas Mountains, which seem to be so high that the heavens rest on them.

He next came on the scene where the hapless Andromeda lay chained to the rocks and, as every proper hero should do, he was able to estimate the situation at a quick glance. He fought the monster, slew it, and claimed Andromeda as his reward.

The family prepared a great banquet, and were about to eat it, when a spurned suitor of Andromeda, Phineus by name, came uninvited with his companions to the banquet hall and made a disturbance. Perseus told his friends to turn their heads; he then pulled the head of Medusa from the bag, exposed it to the unwelcome guests, and thereby changed them all to statues.

Perseus gave the head of Medusa to Minerva, who

LOOKING NORTH

LOOKING SOUTH

This map represents the sky
at the following standard times

DECEMBER 1 at 10 p.m.
DECEMBER 16 at 9 p.m.
JANUARY 1 at 8 p.m.

affixed it to the center of her shield, and in repro-
ductions of this goddess the horrible face of Medusa,
with serpents for hair, can be seen.

Cepheus, Cassiopeia, Andromeda, Perseus, Pegasus, and Cetus, the Sea Monster, are all in the heavens, strewn in December from the North Star almost
to the southern horizon. These constellations are the
illustrations for this epic story.

BEYOND THE UNAIDED EYE

ABOUT seventy-five years ago, a science writer, Garrett P. Serviss, produced two very interesting books — *Astronomy with the Naked Eye* (long out of print) and *Astronomy with an Opera Glass* (reprinted in 1961 in a revised edition by Chatham Press, Montreal). If for no other reason than to sample the flowery style of this nineteenth-century writer, both of these books deserve reading, if they can be found on local library shelves.

Since World War II, a great many relatively inexpensive binoculars and other optical aids have come on the market, either as war surplus items or as Japanese imports. Even the very inexpensive field glasses or so-called "bird glasses" are useful in exploring the sky, although they are not usually of the highest optical quality.

Field glasses ordinarily have low magnifying power — from two to four diameters — and, like opera glasses, use the simple optical system that Galileo employed for his primitive telescopes early in the seventeenth century. Because of this, the field (the amount of sky visible at one time through the glass) is small. If someone offers "binoculars" at a ridiculously low price, through mail-order advertising or otherwise, the prospective buyer should examine the instrument or its illustration to see if the *objective* (the large lens in the front end of each barrel) is directly in front of the *eyepiece*. If so, the instrument is a field glass, not a binocular in which usually excellent optical elements are used. A field glass might cost as little as four dollars; a binocular might possibly be had for as little as fifteen dollars. The word *prism* before the word binocular is a guarantee, because binoculars contain prisms to shorten the over-all tube length, to produce upright unreversed images and to enhance the three-dimensional aspect of objects on land. Some manufacturers of field glasses very unfairly build in bulges on either side, simulating the prism boxes of true binoculars, in a deliberate attempt to mislead the unknowing buyer.

For browsing through the sky, a monocular (one half of a binocular) is likely to be as useful as a binocular, and less than half as expensive, one reason being that there is no Federal excise tax on a monocular! Binoculars can be adjusted for the interocular distance (the separation of the eyes for any individual), but this is a nuisance when more than one person is to use the instrument in an observing session. With a monocular, this inconvenience is avoided because only one eye is used, as with a telescope.

Some advertisers state that their field glasses make objects appear "four times bigger," or "nine times," in order to state a value comparable with the true magnifying powers for binoculars or monoculars, but this is usually dishonest, because they are referring to area, not diameter. The magnifying power of any optical instrument should always be in terms of diameter. Thus, if the power is given as seven (sometimes written as "7x" — seven times), the object viewed will appear seven times as high and seven times as wide as it does to the unaided eye; it will have an apparent area forty-nine times as great as that seen with the unaided eye, but the magnifying power is only seven diameters or seven times.

And, of course, it is silly and misleading for an advertiser to say, "You can see objects fifty miles away!" In the first place, because of the curvature of the earth's surface an object would need to be about 1700 feet above the earth to be seen from the surface (or the observer could be 1700 feet high and see an object on the ground fifty miles away). But with the naked eye you can see the moon, about 240,000 miles away, the sun at a distance of 93,000,000 miles, stars at distances of hundreds of light-years and even, in a clear, dark sky, the galaxy in Andromeda which is more than 2,000,000 light-years distant!

In any optical aid for studying the sky, we are interested in three "powers." One is *light-gathering power* — how much light the objective gathers as compared with the light that enters the pupil of the unaided eye, or the brightness of the image. The second is *resolving power* — how close a pair of stars can be to each other and yet be separated by the instrument, or how small a detail can be seen on the moon or a planet. The third is the *magnifying power*, which was discussed earlier.

The greater the area of the objective (the front lens of a monocular or other instrument), the greater will be the light-gathering power. The action is somewhat like that of a funnel, which has a wide mouth and directs the liquid out through a small hole at the bottom. The objective concentrates all the light that enters it to a small area at the bottom called the *focus*, where the image is viewed by a small magnifier called the eyepiece. The higher the magnifying power, the fainter will appear the surface density of the final image of an extended object such as the moon, a planet or a nebula, because the light delivered by the objective will be spread out over a larger area by the eyepiece. But a star, which even through the largest telescopes appears to be only a tiny point of light, does not become fainter with higher magnifying power. Indeed, it may appear brighter, because the higher power makes the sky appear darker. That is why, even with a small telescope, it is possible to see some of the brighter stars in a clear daylight sky, if the exact place of the object is known and can be spotted.

In complete darkness, the diameter of the pupil of the normal eye will be slightly more than one-quarter inch. But we are almost never in complete darkness for a long-enough period to permit the pupil to open to this maximum aperture. Under a clear nighttime sky, far from city lights, the pupil's diameter may occasionally be as large as one-fifth inch. This is the aperture that admits light to the retina, the sensitive light-perceiving surface on the inside of the rear of the eyeball. But if we look at the moon, the pupil of the eye automatically gets smaller, admitting less light from the bright object.

Let us use some examples. A popular binocular or monocular will be marked 7x50. The 7 is the magnifying power in diameters, the 50 is the aperture or diameter of the objective in *millimeters*. There are close to 25 millimeters in one inch, so 50 millimeters is nearly the equivalent of two inches. This aperture is ten times as great as the diameter of the pupil of the eye under a dark moonless sky away from the glow from city lights, so the area is a hundred times as great, and the objective gathers a hundred times as much light as the eye alone.

The magnifying power is 7x, so if we look at an extended object — moon, planet, nebula — this light will be spread over 49 times as much area as that seen with the naked eye, so the brightness of each part of the area will be *twice* as great — 100 divided by 49. For this reason, these 7x50 binoculars became known as "night glasses" during World War II; the outlines of distant ships or other objects at night could be discerned better with these glasses. (Actually, because of some internal reflections, even with "coated optics," and some absorption by the glass,

the final brightness is somewhat less than twice as great.)

The 7x50 binocular weighs about two and one half pounds. Especially for daytime use by yachtsmen and sports enthusiasts, the 7x35 binocular, weighing about one and one half pounds, is sometimes preferred. On a dark night, the objective diameter of 35 millimeters is about seven times that of the pupil of the eye, so the light-gathering power is 49 times that of the eye. The magnifying power is again 7x, so the area of an extended object is 49 times, and the resulting brightness of the final image is theoretically the same as that seen with the unaided eye.

Other sizes of binoculars commonly for sale are 6x30, 9x50, even 20x50 and 20x70. If you wish to compare the relative light efficiency of these instruments, divide the diameter of the objective in millimeters by the magnifying power, square the result (multiply it by itself, that is), then divide by 25. The result will be the final theoretical brightness of the image as compared with that seen with the unaided eye. Again, however, we must realize that no optical instrument delivers to the eye all of the light that enters the objective, because of losses by reflections and absorptions on the way through.

Earlier, we saw that a star's image cannot be magnified, so the factor we are here interested in is the light-gathering power. A 50-millimeter objective makes a star appear about 100 times as bright as it appears to the unaided eye, so stars only 1/100 as bright as the faintest visible to the unaided eye can be seen through the instrument — that is, about magnitude 10, in practical use. It is generally agreed that only about three thousand stars visible to the unaided eye stand above the horizon at one moment, on the average, and many of these are dimmed and some are invisible, because of the absorption and scattering by the denser air and longer light path near the horizon. Despite the common opposite impression, it is doubtful if more than two thousand stars are visible at one time on a clear, moonless night. The number visible with the aid of a 50-millimeter binocular or monocular, or a 2-inch telescope, would be about 110,000!

Something might be said here about the relative merits of monoculars and binoculars beyond the items of cost and weight. Many people believe that we can see fainter objects with two eyes than with only one. This is not true, as anyone can easily demonstrate by viewing the faintest stars alternately with one eye and with two. Some people have one eye better than the other, of course, and the good eye should always be used. Maybe one eye will see fainter objects than the other, while the other eye may be able to distinguish more sharply the fine details on the moon or the planets. Some people cannot close one eye without squinting with the other, but this difficulty can be overcome by using an eye patch. But a monocular will show stars as faint as those visible in a binocular. That is why astronomers' telescopes are monocular, although sometimes a binocular eyepiece will be used, to ease the strain of keeping one eye closed or of ignoring the image it receives. A binocular will enhance the perspective — the form — of a terrestrial scene, but this factor is of no value in observing the distant objects in the sky.

When using these instruments, it is advisable to prop at least one elbow or a shoulder against a steady support, so the image will not move all around in the field of view. Still better is the scheme of mounting the instrument on a tripod, instead of handholding it. Several manufacturers have designed attachments for this purpose; they are usually for sale in camera shops, along with what are called pan-tilt tripods, which provide for moving the instrument both right and left and up and down as much as desired. For the sky, this is an elegant arrangement, and should be used.

So far, nothing has been said about resolving power, except to define it. Referring to Map No. 10 in this book, and the "Looking West" position, we find a small star immediately above the bright star Vega. A very keen eye, on a very good night, will see this star as double. Under test in the laboratory, the resolving power of the human eye is better than this, but under actual observing conditions the laboratory value is seldom achieved.

Close to the star at the bend of the handle of the Big Dipper is a star easily seen with the unaided eye as separated from it (see, for example, Map No. 5). With a 7x50 or 7x35 monocular, the bright star Mizar and the faint star Alcor will appear to be separated by about two and one-half moon-diameters! And between the two stars, some distance off the line joining them, a third star will be glimpsed. With even a small amateur telescope which provides a magnifying power higher than 7x, the bright star Mizar will be seen to be double. But, if the resolving power of the objective is not great enough, no

amount of magnification will split two stars that are very close together, or separate the stars of a tight cluster, or show small craters on the moon or small details on the planets.

The resolving power of an optical system depends primarily on the diameter of the objective. (It depends also on the effective wave length of the light.) We cannot consider the eye to be a typical optical system in this respect, however, because the grainy structure of the retina sets the limit at about 180 seconds of arc, for an average good eye. The two stars of Epsilon Lyrae, mentioned earlier as being near Vega, are separated by 208 seconds of arc, and a good eye should be able to separate them.

For two equal stars of magnitude 6, just about at the limit of naked-eye visibility under excellent conditions, the resolving power of an objective is usually given in terms of what is called Dawes' Limit — 4.56 seconds of arc (4."56) divided by the aperture (diameter) in inches. So a 2-inch objective can separate two stars that are only about 2.3 seconds of arc apart. But we have seen that the resolving power of the eye is about 180 seconds of arc, so we must magnify enough to make the 2.3 seconds look like 180 seconds; to do that, a magnifying power of about 80x will be necessary.

The components of the star Mizar are about 14 seconds of arc apart. To enable the eye to see them as separate, then, a magnifying power of 13x or 14x is needed to separate them. A popular-size amateur reflecting telescope (to be discussed later) has a primary mirror — its objective — six inches in diameter; according to the Dawes' Limit, the resolving power (for two stars of magnitude 6) should be about 0.76 second of arc. Again assuming the resolving power of the eye to be 180 seconds of arc, the telescope betters this by a factor of about 240 times. This is 40 times the diameter of the objective in inches, just as we found that the magnifying power of 40 times the aperture of a 2-inch objective would realize its resolving power.

Any magnifying power beyond 40 times the aperture of the objective in inches, for any optical instrument used in astronomy, is called "empty magnification." However, a power somewhat higher than this is often used, to make critical objects a little easier to see. We usually set the practical limit of magnification for a telescope at 50 times the aperture in inches. Beyond this, almost nothing will be gained, because the resolving power is determined by the diameter of the objective and the resolving power of the eye, which is in turn determined by the granular structure of the retina.

Beyond this concept of empty magnification, another barrier to unlimited magnifying power exists. This is the atmosphere of the earth, which is always in a state of turbulence because of layers and columns of air at various temperatures. The telescope magnifies this turbulence just as much as it magnifies the object being viewed. This unfortunate phenomenon can be seen as we look over a hot road on a summer day, or over a hot radiator in winter. The distortion of the scene is what is called "bad seeing."

Even in high mountain observatories, established there to get above the denser, more turbulent air, some of the best professional work done visually on lunar or planetary detail has resulted from observers' using large telescopes (to gain resolving and light-gathering powers) with magnifications of 500x or less. Rarely are magnifying powers as high as 1500 used, even on telescopes whose empty magnification might begin at 3000x or 4000x. When an astronomical observatory can be established on the moon, where there is no atmosphere, any telescope can use the power at which empty magnification begins, and we shall see things we have never been able to see before. The 200-inch reflector now on Palomar Mountain could use a power of 10,000x, if it were on the moon; and spots only three or four miles in diameter on Mars could be discerned, as compared with the twenty- to twenty-five-mile areas discernible from the earth.

A large telescope receives light that has passed through a cylinder of turbulent air larger than that through which light passes to a small telescope. For example, the amount of turbulence in the light path for the 200-inch Hale Telescope on Mount Palomar is four times as great as that for the 100-inch Hooker Telescope on Mount Wilson, the atmospheric conditions being the same. As we have seen above, the empty magnification for the 200-inch begins at 10,000x for visual observations, a power that never could be used because of the overpowering turbulence that would distort the image, even on what might be considered a fine night. The virtue of the larger aperture of the 200-inch lies in its higher resolving and light-gathering powers, so fainter objects can be photographed with equal exposures, nearby bright objects can be photographed in a shorter time, and details can be more sharply distinguished (because of higher resolving power).

The emphasis placed by an amateur on magnifying power often leads to disappointment and infrequent use of the instrument. Several eyepieces, giving various magnifying powers, are usually provided with a telescope or can be purchased separately. Beginning with a low power, the observer should examine the image, say, of the moon or Jupiter or Saturn. Then another eyepiece can be used, to see if the higher power shows more. An eyepiece of still higher power can then be used, and so on, until it is found that the image is no better or, indeed, is even worse. Then a lower power can be settled on. For an extended object, as we have seen, higher magnifying powers dilute the surface brightness and increase the distortion caused by atmospheric turbulence. Higher powers can be used on double stars, clusters and star fields of the Milky Way because, while turbulence increases, there is no surface area to grow fainter. The rule is always this: Use the lowest magnifying power that delivers a satisfactory image most of the time. This is the rule followed by professional astronomers and by the serious amateurs who do valuable work or best please the friends and relations whom they invite to look through their telescopes.

What kind of telescope should you buy? There are two main types: *refractors* and *reflectors*. The refractor has a *lens* at the sky end, through which light from a celestial object passes, to be bent or *refracted* to form an image at the *focus* at the lower end of the tube. There the eyepiece is placed to magnify the image. Binoculars and monoculars are small, low-powered refracting telescopes. The reflector has a *concave mirror* as its objective, at the lower end of the tube. Light from the object travels down the tube, then is *reflected* back on itself by the *primary* so that it would normally form an image near the upper end of the tube. Before it reaches the focus, the light in the usual amateur reflector is bent by a flat *secondary* mirror through a right angle to the side of the upper end of the tube, where the eyepiece is located. The observer looks into the side of the upper end of the tube of such a reflecting telescope; the refracting telescope is like a monocular or a "spy-glass" in that the observer looks upward through the eyepiece at the bottom of the tube.

The largest reflecting telescope in the world is the Hale 200-inch on Mount Palomar; the primary mirror, or objective, is 200 inches in diameter, about large enough to park a modern automobile across it! The largest refracting telescope is the 40-inch at the University of Chicago's Yerkes Observatory, located at Williams Bay, Wisconsin.

Always, astronomers have wished to have greater apertures for their telescopes, for greater light-gathering and resolving power. The 200-inch has five times the resolving power and twenty-five times the light-gathering power of the 40-inch.

Because various wave lengths of light (different colors) are not refracted equally, a single-lens objective of a refractor would yield images with colored fringes around them. This is called *chromatic aberration*, which can be largely corrected by making an objective of two lenses of different kinds of glass — the *achromatic* objective. These lenses must be made of glass very clear and free of bubbles; and four accurate surfaces — two for each of the lenses — must be precisely ground and polished. A mirror, on the other hand, reflects all kinds of light identically; besides, only one surface of the primary mirror and one of the secondary must be ground and polished, and the glass behind these surfaces need not be of the fine clear optical quality demanded by a refractor, because the light is reflected from the aluminum-coated front surface and does not pass through the glass. Consequently, a refractor's objective is more expensive than the optics of a reflector of the same size; and refractors, inch for inch, cost much more than reflectors. As a final consideration for professional astronomers, it has been found technically impracticable to make the fine pieces of optical glass for the objectives of telescopes larger than about 40 inches in diameter; supported only at the rim, such a large lens tends to sag from its own weight, while a mirror can be supported all over its back.

About the same time that Isaac Newton, in the seventeenth century, designed and constructed the first of the simple type of reflecting telescope described above, the Frenchman Cassegrain suggested that after the light leaves the primary mirror at the bottom of the tube to be reflected upward, it be reflected back downward by a *convex mirror* to the bottom of the tube, to pass through a hole in the center of the primary to the focus, to be examined by an eyepiece. This Cassegrain type of instrument is arranged for almost all large professional instruments, in addition to the Newtonian system, and many amateurs have built such instruments.

More recently, various other kinds of *catadioptric* telescopes — folded back on themselves as in the Cassegrain, but having thin specially curved lenses in the upper ends of the tubes and mirrors at their bottoms — have been designed and used. They are very fine instruments, testing up excellently in the laboratory and delivering exceedingly sharp images in use on the sky. However, they are several times as costly as a Newtonian instrument of the same size and, except for the advanced amateur with plenty of money, they are not recommended.

Usually, the amateur wants as much telescope of acceptable quality as he can buy for a relatively small amount of money. Without going into all of the details, it can be suggested here that the best buy would be a 6-inch Newtonian reflector, as a start; it can be purchased for $200 and upward, depending on the manufacturer and the number of eyepieces and other accessories provided.

Thousands of hobbyists have made the mirrors for their own telescopes. All the materials necessary for a 6-inch mirror can be purchased for about twelve dollars and a book with all instructions (such as Allyn J. Thompson's *Making Your Own Telescope*) can be had for four dollars; both the materials kits and the book are advertised in the magazine *Sky and Telescope*, mentioned earlier. As one writer put it, after "ten thousand trips around a barrel," the mirror is finished in twenty to forty hours, depending on the skill of the maker. Then it is sent off to have the brilliant aluminum coating administered to the front surface, in a vacuum chamber, at a cost ranging from six dollars upward.

A professional finished 6-inch mirror can be purchased for about fifty dollars; its cost may represent only one fourth or one fifth of that of a complete telescope, ready to operate on the sky. Except for one who is interested more in handiwork than in looking at the heavens, making a mirror is not recommended. Too often, after the mirror is made, the hobbyist will not provide a suitable mounting so it can be used; there are eyepieces to buy and machine work in building the mounting. As a rough guess, perhaps fewer than one in eight of all mirrors made by amateurs has been mounted well enough to encourage the maker to use it. The breed of amateur telescope makers must be sharply distinguished from that of amateur astronomers. Making one's own mirror is a fascinating challenge and yields great satisfaction; but while the mirror is being made it is good to have a usable telescope available, if the interest is truly in exploring the sky.

It is almost essential that the telescope mounting be what is called *equatorial* and that it be equipped with a *driving clock* that keeps the object under observation in the field of view as the earth rotates on its axis, once the telescope is pointed at the object. A battery of several eyepieces should be available, for various magnifying powers, and a *Barlow lens,* which doubles or trebles the power of an eyepiece, is a valuable accessory that can cut down on the number of eyepieces needed. A 6-inch reflector might have eyepieces which, either by themselves or with the amplification of a Barlow lens, yield magnifying powers of about 40, 80, 150, 200, 250 and 300 diameters. Most of the time, a power of 125 to 150 will be considered "high"; on exceptional nights, 200 and 250 might be used; only once in a *great* while — unless the telescope is used in a high, dry country far away from city lights and industrial turbulence — can 300 be used.

There are many handbooks for the telescope user, to tell him where various objects suitable for observation are located and to tell him, for example, that the instrument should never be pointed at the sun without proper auxiliary equipment and instructions. These are advertised in the same magazines that carry notices of telescopes, binoculars and monoculars for sale. Some of the atlases and star charts suggested elsewhere in this book have lists of suitable objects or have them indicated on the maps themselves.

One parting word of caution: Do not expect to see star clusters, nebulae, galaxies and comets through your telescope as you have seen them pictured in books. These are photographs taken through large telescopes; and photographs, except for views of the moon and planets, always reveal more than the eye, using the same telescope, can ever see. The photographic plate or film can accumulate light to build up a fine strong image, over a period of many minutes or even many hours, whereas the human eye can "take" only "snapshots." But a telescope or even a monocular can increase appreciation of the sky and lead to much satisfaction.

THE SUN'S FAMILY

WE LIVE on the surface of a densely populated earth, where, in some of the largest cities, there is a minimum of space in which to move about. This, if terrestrial experiences are taken as representative, would indicate that the universe is an endless sea of space, with all sorts of objects and materials crowded together, swirling around in it. Yet when astronomers survey the universe and plumb its most remote reaches, they discover that the universe is primarily empty space. One of the greatest concentrations of celestial bodies in the universe appears to be right here in the domain of the sun, where nine planets, billions of comets, an unknown number of satellites, including our moon, and a colossal assortment of debris are to be found. It can prudently be said that the thing we have most of in space is space itself.

As an example of the emptiness of space, when we move out from the sun—say sixteen light years, or about a hundred million million miles—there are only fifty-three stars in that tremendous shell of space. One of these fifty-three is the sun. If in some mysterious fashion we could shrink the hundred trillion miles to the diameter of the earth, we would find in this hollow earth fifty-three objects representing the stars and ranging in size from a pea to a tennis ball. There are also planets circling at least one of the objects in this volume of space, but the largest of the planets circling the object representing our sun would be no larger than a poppy seed, with the other planets ranging down to the tiniest grain of salt.

Thus, from a cursory inspection of space around us, we find that the greatest crowding is in the immediate vicinity of the sun—in essence, the sun's family—and it is the planets that we plan principally to explore. Five of the planets are visible to the unaided eye, while the other three require a telescope. We are, of course, ignoring the earth in this list. Some of the planets are among the brightest objects in the sky; others require a large telescope to view them. Other solar-system members, the asteroids, can only be photographed; and still others, the comets, come in for a brief erratic visit with the dominant sun and become provocatively beautiful at their brightest.

To better understand some of the physical features of the solar system it is always helpful to build a model. Models possess the unique quality of shrinking or enlarging realities until they become more readily comprehensible to the layman. For this reason, we will try to model both the sizes and the distances of the major members of the solar system—the planets.

As a focal point, let's choose the Empire State Building in New York City. We will shrink the 864,000-mile

diameter of the sun to about 1,400 feet—the approximate height of that building. Once we have established that where the Empire State Building stands is a 1,400-foot sphere representing the sun, the scale of the system has been determined and we can begin formulating our model.

From the focal point, we will move south and west about 11 miles, to Newark, New Jersey, and there, in a public square, place a five-foot ball. This ball represents the 3,100-mile-diameter Mercury, whose true distance from the sun is about 36,000,000 miles. Next in order of distance, we come to the 7,700-mile-diameter Venus, which is about 67,000,000 miles from the sun. We would represent this by a 12.5-foot sphere in Linden, New Jersey, about 20 miles from New York. Next comes the earth, its 7,900-mile diameter represented by a 12.8-foot sphere in New Brunswick, New Jersey, and its 93,000,000 miles from the sun projects the planetary sphere out 29 miles from New York. The last of the so-called "terrestrial" planets is Mars, 4,200 miles in diameter. Mars would be represented by a 6.8-foot ball at Princeton, 43 miles from New York, which corresponds to the true solar distance of 141,000,000 miles.

Before we move out to the giant planet Jupiter, we must first acknowledge the presence of the asteroids or planetoids that comprise over 100,000 bodies with an enormous range of sizes. The largest is Ceres, 488 miles in diameter, and the smallest must be on the order of tens of feet in diameter. It is generally conceded that an asteroid has to have a diameter of about one mile to be observed visually. For our model, Ceres would be a ball 9.5 inches in diameter. Some asteroids come rather close to earth. Icarus, in 1968, came within 4,000,000 miles of the earth; there are some others that come still closer. Thus, in our model, the bulk of the asteroids would be distributed in a large, thick, doughnutlike ring coming as close to the sun as the earth; in our case, New Brunswick indicates the earth's distance from the sun. Others will move out to Jupiter, which in our model is in Sunbury, Pennsylvania, about 150 miles from New York.

Jupiter represents the first of the major, or "nonterrestrial," planets, for it is the largest planet in the solar system, with an equatorial diameter of about 88,980 miles. Its distance from the sun, 480,000,000 miles, is represented as a 142-foot ball at Sunbury, Pennsylvania, a small town on the Susquehanna River. Beyond Jupiter, at a solar distance of 885,000,000 miles, lies Saturn. This 75,000-mile planet could be represented by a 126-foot ball in Johnstown, Pennsylvania, 276 miles from New York. The rings of Saturn would be about 300 feet in diameter and would have a thickness less than the thickness of this page.

The first planet discovered in recorded history was Uranus, about 1,785,000,000 miles from the sun. With a diameter of about 30,000 miles, it could be represented by a 48-foot ball in Lima, Ohio, 550 miles from New York. Neptune, the next planet in the solar system, would be a ball 45 feet in diameter in Peoria, Illinois, 870 miles from New York; this planet, 28,000 miles in diameter, is 2,797,000,000 miles from the sun. The most distant planet in the solar system is Pluto, discovered in 1930. Because it is so small and so distant, its diameter is not known with any degree of certainty. However, it is believed to be about 1,800 miles in diameter, and its true distance from the sun is about 3,670,000,000 miles. On our model, it would be a 3-foot ball in Kansas City, Missouri, 1,150 miles from New York.

On this scale, the nearest star, Alpha Centauri, could be represented by a sphere 1,500 feet in diameter and at a distance of roughly 32 times the moon's distance—about 8,000,000 miles—from New York.

There still remains one class of objects which must be integrated into our model. These are the comets, those magnificently evanescent objects that move in to circle the sun and then move out again perhaps not to return again until thousands of years have elapsed. As long as the origin of the solar system continues to be a subject of controversial debate, the origin of comets too will be shrouded in unfathomable mystery. Currently, it is believed that contained in a shell surrounding the sun and at a distance of about two light years are billions of isolated mountains of ice-coated gravel in a wide spectrum of sizes, and that these mountains travel in roughly circular orbits around the sun, with periods on the order of millions of years. Occasionally, the passage of a distant star will disturb a comet and make it swing in toward the sun. As it approaches the sun, the hydrogen ices—methane, ammonia and water—will be vaporized by the sun, whose short-wave radiations will make the gases fluoresce. Radiation pressure and the pressure of the solar wind (composed of solar particles) will make the gases and particles stream away from the sun to create the enormously long tail, which makes the comet appear as one of the most spectacular celestial phenomena that man can behold.

Before leaving the overview of the solar system, we must remember that the motion of any celestial object around another is in an ellipse. In the case of the solar system, the sun is at one of the foci of the ellipse. This means that because of elongated planetary orbits, the distances from the sun have been given as an average, or mean, distance. In reality, some of the planetary orbits are quite elongated, while others are almost circular.

These, then, are the objects which abound in this part of space and make this part of space one of the most congested in the universe. Now let's explore the intimate details of the planets of the solar system.

MERCURY

If from the earth we cannot determine the physical characteristics of a planet, then let's send a spacecraft to provide a close-up view. This is the Space Age and we are capable of launching probes to the distant planets; so, why not do it? This is precisely what has been done. Planetary spacecraft have been launched to Venus, Mars, Jupiter, and Saturn, and one has passed Saturn and is on its way to Uranus and hopefully Neptune. Why not launch a spacecraft to explore one of the most difficult planets to observe? Why not, indeed? For this reason, Mariner 10 was launched November 3, 1973, on a two-planet mission. It passed Venus on February 5, 1974, and on March 29, 1974, passed within 450 miles of the surface of Mercury. Astronomers learned more from this single planetary spacecraft than from all other observations made, going back to the time when ancient watchers of the sky saw Mercury as a morning or evening star in the early dawn or evening twilight.

On this first pass, more than eight hundred pictures were relayed back to the earth, some of which disclosed features no more than a few hundred yards in diameter. Almost two hundred pictures resulted from the second and final operating pass.

Mercury is the planet closest to the sun and moves in the second most eccentric orbit of all the planets. At aphelion, its greatest distance from the sun, it is 43 million miles distant. At perihelion, its closest approach to the sun, it is but 28 million miles distant. Its year is but 88 days in length, and the synodic period, the time it takes the planet to go from between the earth and the sun back to the same position, is 116 days.

Because it is so close to the sun, Mercury can be seen in the morning and evening skies, but never more than 28 degrees from the sun. Thus, it can be observed close to the horizon after sunset or before sunrise, and the presence of even thin clouds will obscure it. When it is observed in the twilight, the duration of observation is very short. In the late nineteenth century, some astronomers tried to observe it in full daylight. It was apparent that their daylight observations did little to help determine the physical characteristics of the planet. We see less on Mercury with our giant telescopes than we see on the moon with our naked eyes.

In the early nineteenth century, the day on Mercury was calculated to be slightly in excess of 24 hours. By the late nineteenth century, observations indicated that the planet always showed the same side toward the sun. This meant that the day and the year were equal in length, and as the year on Mercury is 88 earth days, it spins on its axis in the same period of time. With the advent of radio astronomy, radar observations in the middle 1960s showed a day of 58.6 earth days, which not only was unexpected but came as a slight shock. Astronomers have pointed out that with the day 58.6 earth days in length and the year 88 earth days in length, the ratio of the Mercurian day to the Mercurian year is 3 to 2. That means that Mercury experiences three days in two years. The precise cause of this curious simple ratio is unknown, but it is believed that the sun is pulling on a bulge in the planet's distribution of mass.

Mercury has a diameter of 3,100 miles, with a mass about 5 percent that of the earth. These figures yield a density about 5.5 times that of water, which is about the same as the earth's. Our estimates of its diameter are difficult to make with certainty. Similarly, because it has no satellite, our knowledge of its mass is somewhat uncertain. If Mercury has a satellite, it must have a diameter under 3 miles. Any change upward in these figures means that Mercury could become the planet with the highest density. Astronomers suggest that its proximity to the sun, with the sun driving off the lighter elements, could account for the high density.

When Mariner 10 flew past Mercury, it disclosed that, contrary to popular belief, the planet surface has a magnetic field with a strength about 1 percent that of the earth's surface. In the case of the earth, scientists believe that its magnetic field is due to a dynamo effect from the rotation of molten materials within the core of the earth. However, in the case of Venus and Mercury, the spin rate is so slow that no magnetic field is expected. Yet Mercury does possess a weak magnetic field of about 800 to 900 gammas, compared with 30,000 gammas at the earth's equator. Even this weak a magnetic field dictates an iron core, and from the high density of the planet it is probable that there is an iron core about 2,000 miles in diameter surmounted by a rock mantle about 500 miles thick. Mariner 10 also disclosed that the sun-side temperature ranged from 560 to 800 degrees Fahrenheit, with a night temperature range of minus 300 degrees. This represents a temperature range of about 1,000 degrees Fahrenheit, which could mean that the planet's surface may be covered by a layer of porous material with a low thermal conductivity.

The proximity of Mercury to the sun and its rather high surface temperature and low gravitational field (because of its small size) indicate that there should be a minimal atmosphere on the planet and that hydrogen, the principal component of the solar wind, should constitute the atmosphere. But Mariner 10 disclosed an appreciable atmosphere rich in helium. The real surprise came when it was found that sodium was also a major constituent of the atmosphere. It is suspected that sodium is of meteoric origin.

Television pictures of the desolate surface of Mercury revealed striking resemblances to features on both the moon and Mars. The surface is rough, cratered, and scarred by the impact of space debris on its way toward a closer orbit of the sun. There are craters which appear heavily eroded, even though erosion processes depend on an atmosphere with accompanying winds. Some of the craters have central peaks typical of those found in lunar craters. It appears that there is a greater variety of crater forms on Mercury than is found on the moon. They are definitely impact craters, in which material is blasted out of the crater, with a significant portion falling back to create the central peak. The craters are of varying ages; some are found on the rims of larger craters or on the floors inside them. Lava-filled craters or flows similar to some lunar features can also be seen. Ray systems, similar to those on the moon, have also been found on Mercury, and the mechanism for their formation is again the impact of space debris, spraying the surface with subsurface materials. The largest feature is the Caloris Basin—a circular wall over a mile high and about 800 miles in diameter. There are rills, deep valleys, and fault lines to be seen on the surface. A 300-mile-long ridge over 1½ miles high has been seen. Such a feature on earth is usually attributed to the warping of the surface by deep internal processes. Mercury is so small that its interior should have been stabilized, but it appears to have been chemically differentiated at some time in its history. Both Mars and the moon are heavily cratered, and indeed, it may be that most of the planets—if there were minimal or no erosion—would exhibit these features.

VENUS

The most brilliant object in our sky, aside from the sun and the moon, is Venus. The early Greeks did not believe it was the same brilliant object that they saw east and west of the sun, so they called it Phosphorus as a "morning star" and Hesperus as an "evening star." It was Pythagoras, in the sixth century B.C., who apparently was first to recognize the morning and evening star as the same planet.

As with Mercury, the orbit of Venus lies inside the orbit of the earth, and thus Venus can never get very far from the sun. Its greatest distance from the sun (elongation) is 47 degrees. Because of this limited angular distance, it can never set more than three hours after earth's sunset or rise more than three hours before our sunrise. Venus can approach within 25 million miles of the earth, and at such times it is our closest planetary neighbor.

Venus has been called the planet of mystery, for we had never seen its surface. The Soviet Union with its Venera spacecraft has landed probes on the surface, and startling color pictures of the surface have been relayed to earth. The Venusian rocks apparently have a density and composition of natural radioactive elements similar to those found in terrestrial basalts.

NASA's Pioneer Venus 1 and 2 discovered that Venus is covered with a permanent, dense carbon dioxide atmosphere. However, hydrogen, helium, carbon, oxygen, and sulfur compounds such as carbonyl sulfide, hydrogen sulfide, and sulfuric acid are also present to press down on the Venusian surface about 40 miles below with a pressure about 90 times that of our atmosphere. It is this thick, dense atmosphere which accounts for the brightness of the planet. Maximum brightness, which takes place when Venus is 39 degrees from the sun, occurs thirty-six days before or after Venus is at inferior conjunction—that is, when the planet lies between the earth and the sun. At maximum brightness, Venus is an exquisitely beautiful object 10,000 times brighter than the faintest star visible to the naked eye.

Because it does become so bright, Venus is the only planet that can be seen with the unaided eye in broad daylight. At its brightest when the sky is dark, it can actually cast shadows. However, if one wants to see Venus in daylight, one has to know precisely where to look. In the Observatory of The Franklin Institute, one would sight up along the barrel of the 10-inch Zeiss refractor and thus direct one's vision to a restricted part of the sky where one would see the planet. I have done this many times, and it comes almost as a shock to see this bright object in the daytime sky.

In the early days of the telescope, astronomers thought they could distinguish markings on the surface of Venus, and they came to the conclusion that it rotated on its axis once in about 24 hours. Then the figure 225 days was deduced with the indication that Venus always keeps the same side toward the sun. However, since the advent of the colossal radio telescopes, astronomers have established that Venus rotates with a period, or day, of 243 earth days, which is longer than the 225 days of its year. This is a startling disclosure, for all the planets in the solar system rotate in the same direction as that of their motion around the sun. Venus represents the only planet in the solar system that rotates in a retrograde direction—that is, opposite to its motion around the sun. (We will not include Uranus in this discussion for reasons which will become apparent later.) As the year on Venus is 225 earth days and its day is 243 earth days, the sun rises for Venus every 117 days. Contrary to what is normal, the sun on Venus rises in the west and sets in the east.

Results of space probes indicate that Venus has a magnetic field less than 0.1 percent that of the earth and no radiation belts. These observations are compatible with a slowly spinning planet. There is a curious relationship between the spin of Venus and the relative orbital motion of Venus and the earth. As the day on Venus is 243 earth days, at every inferior conjunction, when Venus lies directly between the earth and the sun, Venus presents the same face to the earth. This indicates that in some as yet inexplicable fashion the earth influences the spin of Venus.

Venus has been the subject of many planetary probes. In 1962 the United States launched Mariner 11, which flew by the planet on December 14, passing within 22,000 miles of the planet. A temperature probe scanned the planet and recorded a surface temperature of 600 degrees F. Other probes, including Mariner 10

and the Soviet spacecraft Venera 7, which landed on the Venusian surface, transmitted information from the surface indicating a lead-melting temperature of 900 degrees F. Because Venus possesses a heavily insulating atmosphere and rotates so slowly, this temperature is constant for both day and night sides. Mariner 10, which passed Venus on February 5, 1974, relayed back to the earth pictures in ultraviolet light showing a large, irregular dark pattern in the planet's equatorial region. This "great Venusian eye," as it was called, is roughly 4,200 by 1,200 miles. It is an atmospheric feature, and scientists indicate that they can see down into the murky atmosphere to a depth of about 37 miles above the planet's surface. They also studied the motions of the clouds around Venus, and these appeared to be moving with a speed of about 220 miles per hour, fifty times the spin rate of the planet.

The great Venusian eye is not the only feature that is prominent. Television pictures reveal that the cloud cover is highly structured. Apparently, solar heating generates convective currents that split the atmospheric flow into a Y-shaped configuration which sends the two branches spiraling out toward the polar regions. The speed of the middle cloud layers increases to 450 miles per hour as they approach the poles. Apparently, the circulation of the atmosphere from the equator to the poles is opposite to that on the earth, where the flow is from the poles to the equator.

While it is uncertain how far into the atmosphere the pictures penetrate, it appears that at least three distinct layers have been identified. There are a multiple haze layer 44 miles high, a second layer at an altitude of about 37 miles, and a third layer ranging between 18 and 30 miles above the surface.

The composition of the lower atmosphere has been determined, in terms of the percentage of molecules of the gases, as approximately 96.5 percent carbon dioxide, 3.2 percent molecular hydrogen, and traces of water, molecular oxygen, and argon. However, there is also present a significant supply of sulfur, giving rise to sulfur compounds (principally sulfur dioxide), which changes with altitude. Above 12 miles most of the sulfur compounds disappear and only sulfur dioxide persists.

Measurements by the neutral mass spectrometer on the Pioneer spacecraft detected 100 times as much deuterium (heavy hydrogen) as hydrogen in the Venusian atmosphere. Some scientists speculate that when the solar system was very young, say 4.1 billion years ago, Venus had oceans of water containing approximately a few tenths of one percent of the earth's present ocean volume. But the greenhouse effect heated the planet and vaporized the water, permitting the lightweight hydrogen to escape.

Scientists have discovered that the ionosphere surrounding Venus is dramatically dynamic, unlike the rather stable ionosphere of the earth. It can form and disappear in a 24-hour period. Large electromagnetic bursts have been recorded which are ascribed to lightning.

We know very little about the surface characteristics of the planet Venus except for the temperature, which was verified by the Pioneer Venus "Day" probe, which transmitted from the surface for more than 67 minutes after impact. Scientists manning huge radio telescopes have been trying to fill in the gaps in our knowledge from return signals from the planet. Preliminary indications are that the planet surface is, perhaps, as rough as that of the moon. The radar mappers on the Pioneer Venus Orbiter disclosed a long rift valley with walls that stand about 4 miles above the adjacent dusty areas. It is confirmed that Venus is a dry, hot, desolate world, with no probability of any life forms existing. These radar mappers revealed the presence of a huge 2,000-by-1,000-mile plateau with three groups of mountains bordering the plateau. A mammoth peak, unofficially named Maxwell, rises about 8 miles above what is the equivalent of sea level on the earth. Also revealed was the largest known canyon in the solar system, at least 900 miles long, 175 miles wide, and 3 miles deep.

Data from photographs from the Soviet Venera 9 and Venera 10 lander spacecraft portray a rocky surface belonging to a young, evolving planet which, despite the heavy cloud cover, is as bright as the earth on a cloudy day in summer. The stones in the photograph have sharp edges and rounded edges, indicating tectonic (mountain-building) activity. They range in size from the largest, with an area of 12 square yards, to pebbles just visible to the eye. While dust is present on the surface, it does not blow about as on Mars, for wind speeds near the surface are extremely slow.

Radar observations made at the Arecibo Observatory in Puerto Rico disclosed numerous craters, some 200 miles in diameter. Most of these craters have prominent central peaks, indicating that in the past large meteors impacted the planet. However, other observations point to considerable past volcanic activity which gave rise to dark craters that resemble the caldera of volcanoes. There is evidence of two major, currently active volcanic areas. These may provide the principal vents for the planet's internal heat. Further evidence of volcanic activity is deduced from the presence of carbon monoxide, water, and nitrogen.

There is an apparent scarcity of small craters on the surface; this can be rationalized by an assumption that the small meteoroids which could give rise to small craters would burn up in the atmosphere. The hot, dense atmosphere would be a powerful erosion agent if the surface winds were strong enough to sandblast the surface. The pictures relayed to earth disclosed a mixture of rounded and sharp rocks which constitutes the signature of a geologically young landscape. Other areas showing large flat boulders, indicative of a geologically older surface, show that erosion does occur even though there is a lack of water and possibly insignificant wind-borne dust.

Venus moves in a practically circular orbit at a mean distance of 67 million miles from the sun. The eccentricity of the orbit is so small that the planet's distance from the sun varies by only 1 million miles. The year on Venus is 225 earth days, but its synodic period—that is, from the time when Venus is at, say, greatest eastern elongation to the next time it is at greatest eastern elongation—is 584 earth days, or about 19 months. This is why when Venus is at its brightest it catches some people by surprise; they do not remember it. If Venus is brightest following the sun in the winter, the next time it will be brightest following the sun will be in summer. After about 3 earth years it comes back to being brightest in the winter. Apparently, the memory span of the casual observer breaks down over this approximately 3-year period.

Venus has been called the earth's twin, because its measured diameter (at the top of the deep atmosphere) is about 7,700 miles—about 200 miles less than that of the earth. Its mass is 0.87 times that of the earth, and its density is 5.1 times that of water. This high density suggests a dense core somewhat like the earth's nickel-iron core.

Telescopically, Venus is a remarkable object because of its ever-changing perceived size. When it is behind the sun, it is 160 million miles from the earth; when it lies between the earth and the sun, it is only 25 million miles from the earth. The ratio of these two distances is about 6 to 1. When we see Venus in full phase, behind the sun, it is a small body. When we observe Venus coming between the earth and sun, it is six times as large but seen as a thin crescent.

MARS

Mars once represented the most intriguing planet in the solar system because of the incredible volume of speculation that surrounded it for almost a century. Ever since 1877, when the pioneering observations of the Italian astronomer Giovanni Schiaparelli disclosed the apparent presence of faint, evanescent straight-line markings at the critical limit of visibility, Mars has been the object of more concentrated telescopic study and research than any other planet. Schiaparelli termed the streaks *canali*, an Italian word which really means channels but freely translates into English as "canals." Until the advent of the successful Mariner probes to Mars, canals remained a subject of bizarre and at times acrimonious controversy. Today, we know from high-resolution photographs of Mars taken from orbiting spacecraft and the Viking landers that no canals exist. There are surface markings on Mars similar to those on earth and the moon. Some markings are strictly Martian, with no counterpart on either earth or moon. In other words, Mars looks like Mars.

Observations from the earth had convinced Percival

Lowell that there were high mountains on Mars, but because these ran counter to his theory of canals, he ignored his own observational evidence. Pictures relayed back by Mariner 9 indicated that there are indeed high mountains on Mars, with a surprising difference in elevation of 10 miles or more from the top of the mountains to the plains below.

Volcanic craters too are to be seen. One majestic volcano, Nix Olympica, is 500 miles in diameter. In addition to volcanic craters, there is a tremendous number of impact craters which are undoubtedly older than the volcanic ones. Some astronomers have suggested that the time of cratering activity on Mars lies somewhere between that of the moon and that of Mercury, which means going back billions of years.

Pictures disclose a vast array of linear ridges and grooves, one of which is an enormous rift valley in a region called Coprates. It is 50 miles wide, a mile deep, and crosses almost one-quarter the circumference of the planet near its equator. Sinuous channels too, apparently carved by ancient rivers, are visible. There are literally hundreds of dry riverbeds to be seen, indicating a vast flood of swiftly running water in the past. This gives rise to an apparent paradox. If water was responsible for the innumerable, branching, streamlike tributaries, what became of this water?

Like the other planets, Mars spins on its axis with a day of 24 hours, 37 minutes, 22.6679 seconds in earth units. Because the Dutch astronomer Christian Huygens on November 28, 1659, at 7 P.M., made a drawing of Mars on which one of the most prominent markings, the Syrtis Major, was shown, we have determined the day of Mars with an accuracy of 1/10,000 of a second! Apparently, we know the length of the Martian day with greater accuracy than that of the terrestrial day.

The axis of Mars is tipped to its path around the sun by about 25 degrees 10 minutes. Therefore Mars, like the earth, has seasons. However, because of its highly eccentric, or elongated, orbit, the seasons are not of equal length. The southern summer or northern winter is 160 days long. The southern autumn or northern spring is 199 days long. The southern winter or northern summer is 182 days long, and the southern spring or northern autumn is 146 days long. These are measured in 24-hour days.

The planet has a fairly eccentric orbit. Its closest approach to the sun (perihelion) is 129,000,000 miles, while at its most distant position (aphelion) Mars is 154,000,000 miles from the sun. This means that Mars, at its most favorable opposition, can come as close to the earth as 35,600,000 miles, while at its most unfavorable opposition, it is 62,900,000 miles distant. The favorable oppositions occur at intervals of 15 or 17 years. As a point of interest, it is this significant variation in the Mars-sun distance which permitted Johann Kepler to derive his exquisitely simple laws of planetary motion. The discrepancies were so unacceptably large that he had to postulate an elliptical path for Mars in order to have his theory fit the observations.

The rarefied atmosphere of Mars is composed primarily of carbon dioxide, with an insignificant amount of water also present. The remainder is mostly nitrogen, with traces of both oxygen and argon present. The atmospheric pressure on the surface is about 1 percent that of the earth.

Violent dust storms are seen in the planet, and in the case of Mariner 9 for the first few weeks after its arrival at Mars, the turbulent dust storms were so dense and severe that it was impossible to see the surface. Terrestrial observations provide a history of major dust storms on Mars, and it has been determined that they usually occur when Mars is at perihelion. These global dust storms come whirling out of the middle southern latitudes of Mars to mask the entire planet. This occurred in 1971 when Mariner 9 arrived at Mars to find it shrouded in dust, and in 1977 the Viking Orbiter was able to photograph and follow the detailed progress of the dust storms.

Because of the lack of an atmosphere, the temperature variations on the planet are rather severe. The noon temperature at the subsolar point can reach 80 degrees F., while at the same spot the night temperature can plummet to 150 degrees below zero F. At the poles the temperature is so low that carbon dioxide is frozen out of the sparse polar air to form elevations on the polar caps.

Mars, as seen in a telescope or with the naked eye, is a magnificently bright object and is dominated by a rusty, reddish color. Photographs by the Viking landers disclosed that indeed, the boulders, rocks, and fine grains which compose the surface are reddish, and only a rare surface patch will not be vividly reddened. Analysis of the regolith, or top surface, indicates that about 5 percent comprises magnetic materials—that is, oxidized iron (which accounts for the rusty color). Minor constituents of the surface are silicon, magnesium, aluminum, sulfur, calcium, and titanium. Apparently iron—highly oxidized—is present in many forms, such as hematite, magnetite (the terrestrial lodestone), maghemite, and geothite.

In the pictures relayed back by the Viking landers, the Martian sky also appeared to range in color from an orange cream to pink. The color is produced by molecules and suspended dust particles that scatter sunlight incident on the atmosphere from above and reflected from the reddish surface below. Surface particles are of such size as to preferentially absorb blue light and scatter the red light. To do this, the dust particles on the surface must be very tiny—about 25,000 of them standing shoulder to shoulder would just cover 1 inch. On earth, red skies can also be seen at times, but only after a major volcanic eruption which laces the sky with fine particles or when viewed in a violent dust storm on a desert.

Mariner 9 also succeeded in transmitting to earth pictures of the two tiny satellites of Mars. Phobos, like the planet itself, is very heavily cratered and looks like a large, dimpled potato. The fact that it was cratered indicated that the irregularly shaped satellite possesses considerable structural strength and is probably billions of years old. As photographed by Viking 1, it was found to have a top-to-bottom dimension of 12 miles, while in the other direction it ranged from 13 to 17 miles. The other satellite, Deimos, is about half the diameter of Phobos and is the more distant one. The two satellites reflect little light and are among the darkest objects in the solar system. In all probability, they are covered by a fine layer of dust.

The periods of revolution of these two satellites around Mars are the shortest in the solar system. Phobos has a period less than one-third the Martian day. Thus, it moves faster than the surface of the rotating planet to rise in the west and set in the east. It is above the horizon for but 4¼ hours. Deimos has a period of about 30 hours, and this satellite appears to hang almost stationary in the Martian sky. It rises in the east and stays above the horizon more than 60 hours before setting in the west.

Mars is a relatively small planet, with a diameter of 4,200 miles and a mass about one-tenth that of the earth. The resulting density is about four times that of water. The smaller diameter combined with its small mass gives Mars a pull of gravity only 38 percent that of the earth. A 200-pound man on the earth would but weigh 76 pounds on Mars.

The time it takes to circle the sun is 687 days, but its synodic period—that is, the time it takes to go from meridian in the terrestrial night sky back to the meridian—is 780 days. This makes Mars a brilliant evening star about once every 2 years.

The one characteristic of Mars that has fascinated astronomers since the illusive surface markings were first studied was the intriguing and alluring possibility that life of some sort might exist on this planet. They reasoned that the temperatures are not too severe; there is a thin atmosphere, though the amount of oxygen is minuscule; and there is definitely water on the planet surface and in the atmosphere. Could some sort of indigenous life be found on this planet? A partial answer is revealed by the Mariner photographs. There is absolutely no evidence that any form of life exists on the planet. But then, from earth orbit, in only one instance has there been evidence that life exists on earth—and in this case, the astronauts were only 150 miles above the surface and not 35,000,000 miles away. That there was a significant abundance of water on Mars in the past means that life could have been possible. Even if physical conditions were severe, the time frame is so long that with previous favorable conditions it is conceivable that mutations could have arisen to survive the severe conditions. Could survivors from this life be found?

The only resolution to this provocative question is to put a precocious "lander" on the Martian surface from an orbiting Mars spacecraft and have that lander perform biological experiments seeking traces of life on the surface. The Soviet Union unsuccessfully attempted this on seven occasions.

The Viking 1 and 2 Mars orbiters and landers were eminently successful, and the results of their explorations indicated a complete absence of life on Mars. In the words of one scientist, "Mars is deader than a doornail." This means that the earth has evolved as a surprisingly unique planet—the sole member of the solar system that can support a recognizable life. However, if and when man gets to Mars, he will be able to live there. It is not an implacably hostile planet.

Water is an absolute necessity for the beginning and propagation of life as we know it. Thus, to justify a search for life there had to be some assurance that water was present. The results of Mariner 9 revealed that in the past Mars possessed vast quantities of surface water and more atmosphere and more erosive activity than are present today. Mariner 9 revealed many features that resemble dry riverbeds. These channels grow wider and deeper as they run downhill, with many tributaries running into them, and in turn the channels appear to empty into broad, flat plains. The only medium that can account for the carving of these features is water, and in rather copious quantities to give rise to this catastrophic flooding. Today one poses the question "Where did this water go?"

It is conceivable that the water is permanently frozen beneath the surface as a permafrost, for there is a relatively small amount in the atmosphere, and some of the water resides in the polar ice caps. The caps we see expanding during the winter are composed of carbon dioxide, which dissipates in the spring and summer, revealing an ice cap of water. In 1979 the Viking cameras revealed a new thin layer of water frost on the surface. It is believed that dust particles pick up ice particles to which carbon dioxide freezes, making the particles fall to the surface to account for this extraordinarily thin frost layer.

Scientists of rare competence speculate that several billion years ago Mars had a much denser atmosphere and higher temperatures and was subject to intense volcanic activity. The early atmosphere might have contained enough water to form a layer several yards deep over the entire surface of the planet. These conditions are adequate to explain the evidence of water flow, with its violent scouring of the surface. Today, only a tiny fraction of this water remains on the surface or in the atmosphere; the rest is probably locked in a permafrost beneath the surface. This was, it is believed, confirmed by the trench-digging activities of the Viking 1 lander. The shape of the trench indicates that Martian soil is fine-grained and about as cohesive as *wet* beach sand or good farming soil on earth.

Having established that physical conditions could at one time in the past have favored a form of life on Mars, we must confront the results of the physical tests. They were completely negative, giving rise to the expression "deader than a doornail."

The imaginative Viking lander experiments were designed around a carbon-based life which, like that on the earth, can manifest itself as a creature the size of a bear or a bacterium. Thus, the instruments on the lander were designed to detect Martian microbes or similar creatures on the surface automatically in an unprecedented distant and remotely controlled laboratory.

One experiment depended on plants absorbing carbon dioxide from the surrounding atmosphere and converting it into organic compounds. Thus, if radioactive carbon dioxide was added to the atmosphere, the plants would absorb the carbon dioxide, and the radioactivity of compounds in the plants would yield a clue as to whether they were living organisms. The results were negative.

The second experiment was a gas-exchange experiment, in that animals give off carbon dioxide and plants yield oxygen and both exhale water. Nutrients were added to the Martian soil, and this was continuously monitored for biological activity. The results were negative.

The third experiment was based on the premise that terrestrial animals consume organic compounds and give off carbon dioxide. The nutrients used were radioactive, and the experiment was monitored to see if this food was consumed and radioactive carbon dioxide released. The results were negative.

While the results are all considered negative, some form of exciting activity in the Martian soil, which is as yet unexplained, was detected in the experiments. While there is an extremely remote possibility that a form of life is at the basis of this activity, it is possible that an unknown chemical reaction could also account for the activity.

JUPITER

For over three and a half centuries, Jupiter has evolved as one of the most fascinating planets in the solar system. It certainly is the largest and most massive; its mass is considerably more than twice that of the other planets combined. In January 1610, Galileo, with his primitive 30-power telescope, discovered that four moons revolved around the giant planet, simulating a miniature solar system.

Concentrated study of the cloud belts that we see as the visible surface was undertaken telescopically. From careful observation of the motion of markings in the cloud belts, the rotation period of the planet was determined. Jupiter is not only the most massive and largest of the planets, but it has the shortest period of rotation, or day. It spins on its axis in somewhat less than 10 earth hours. The reason for the lack of precision in the determination of the Jovian day is that its rotation, like that of the sun, is a function of the latitude. At the equator Jupiter rotates in 9 hours 50 minutes, while near the poles the rotation period is 9 hours 56 minutes. The year on Jupiter is 11.86 earth years long.

Because of this rapid spin, the planet is flattened by about 9 percent at the poles. The degree of oblateness is such that astronomers conclude that the seething turbulent atmosphere and liquid shell beneath it must be literally thousands of miles deep. As determined by Voyager 1, the equatorial diameter is 88,980 miles and the polar diameter is 83,264 miles.

There is considerable speculation on the internal structure of the planet. However, a theoretical cross section of the planet can be derived. Jupiter probably has a relatively small molten iron-silicate core about 16,000 miles in diameter. This is surrounded by a metallic hydrogen layer 57,000 miles in diameter. In turn this is overlaid with a shell of liquid hydrogen 16,400 miles thick. Finally there is an atmosphere about 600 miles thick which covers the liquid hydrogen. Thus, Jupiter possesses no solid surface. It is simply a swiftly spinning ball of liquid and gas.

The spectroscope, in the hands of skillful astrophysicists, has successfully disclosed methane and ammonia in the cloud bands of Jupiter. Traces of ethane and acetylene are also present. While it may be presumed that carbon, nitrogen, and oxygen are present too, the elements found in greatest concentration (about 99 percent) are hydrogen and helium. Because of the overwhelming presence of hydrogen, in all probability oxygen has combined with hydrogen to create water; carbon with hydrogen yields methane; and nitrogen combined with hydrogen yields ammonia. Helium exists as a free gas, because it is completely inert and does not combine with other elements. There are about 11 percent as many helium atoms as hydrogen molecules in the upper atmosphere.

Cloud belts on Jupiter are both light and dark. The light bands are called zones; the dark brown bands, belts. The rapid spinning of the planet appears to create and maintain this complex system of bands. The light zones appear to be at higher altitudes and to have cooler temperatures than the dark belts.

The size of Jupiter and the highly reflective properties of its cloud cover make this planet, with the moon excepted, the second-brightest object in the night sky. Venus, to be sure, is the brightest; but Venus, as we have discovered, is never seen more than three hours after sunset or before sunrise.

The cloud bands on Jupiter are continually changing their appearance. Buried in one of the cloud bands is the Great Red Spot which may have been first discovered by Giovanni Cassini in 1664 but gained prominence only in the latter part of the nineteenth century. Today, it is considered a permanent feature of the planet. This oval spot has a width of 8,700 miles and varies in length from

18,600 to 24,800 miles. It appears to move forward and back in the atmosphere with a period of about 90 earth days. How do scientists know this? The answer is that they have determined a fixed-coordinate system which is locked—or, really, rooted—in the magnetic field of the planet. In addition to this 90-day oscillation, the Great Red Spot is drifting westward at a rate of 2.6 degrees per day with respect to the fixed-coordinate system. Temperature readings at this spot indicate that the temperature above the Great Red Spot is about 10 degrees F. colder than the surrounding regions. Voyager's infrared instruments have detected hydrocarbons and phosphine—a compound of phosphorus and hydrogen. Phosphine probably comes from below the surface of the spot to rise to the cloud tops. There, ultraviolet radiation from the sun breaks up the compound to release red phosphorus, which provides the color of the Great Red Spot.

Voyager 1 photographs of Juipter reveal a white spot (first noticed in 1939-1940) and a brown oval in the vicinity of the Great Red Spot. Streams of turbulent and chaotic clouds eddy around the Great Red Spot. Inside the Great Red Spot there is relative calm. The blue patches in this area are deeper-lying layers seen through upper-level clearances. The entire area around the Great Red Spot represents awesome, colored, spinning whirlpools in the Jovian atmosphere.

There is a challenging mystery surrounding the heat balance of this planet. Because we know how far Jupiter is from the sun, we can compute how much energy it receives. The astronomer, with his exquisitely delicate instruments, can also measure the energy it emits. Observations indicate that the temperature near the top of the atmosphere should be about 220 degrees below zero F. Actual measurements indicate that it is 200 degrees below zero F. Pioneer 10 measurements of the energy emitted by Jupiter indicate that the planet emits 2.5 times as much as expected from reflected solar radiation. The cause of this discrepancy is believed to be an internal source of heat filtering up to the cloud surface from below the atmosphere of Jupiter. The precise manner in which this heat is generated is a matter of considerable speculation.

Computations by some theoreticians of the central temperature of Jupiter range from 50,000 degrees F. to the highest estimate of about 1,000,000 degrees F. This latter is roughly 10 percent of the temperature needed to initiate thermonuclear or stellar reactions. Thus, unless there is some other overtone of the thermonuclear process of which we are completely unaware, the excess energy cannot come from nuclear reactions. Where, then, does the energy come from? The greatest reservoir of energy in the universe is gravitation. It has been determined that if Jupiter were to shrink roughly 1/25 of an inch a year (a completely undetectable amount), the energy produced by gravitational contraction could ac-

count for the excess energy emitted by the great planet.

There is a third possible source of internal energy. Some scientists speculate that the planet still possesses some residual heat remaining from the primordial heat generated when the planet coalesced out of the solar nebula. This speculation, however, is made with little confidence.

Jupiter has 14 or possibly 15 moons, of which the Galilean moons are the four giants. The fourteenth moon was discovered by two members of the Voyager II Imaging Science team on pictures taken July 8, 1979. The fifteenth moon may have been discovered by Charles T. Kowal, who discovered moon number 13 in 1974. The Galilean moons are large and move in the plane of the planet's equator, which means that when they pass in front of the planet they cast round black shadows on the cloud atmosphere which can be easily seen in a telescope. If one were on the cloud surface at this point, one could see the sun eclipsed. As the moons orbit the planet, they can pass into the long Jovian shadow, where they are eclipsed. The outer satellites move in orbits highly inclined (25 to 28 degrees from the equatorial plane), with the four outermost pursuing retrograde paths.

Probably the most exciting and surprising result of the Voyager spacecraft has been the revelation of an overwhelming plethora of new and totally unexpected surface features of the Galilean moons. Of this plethora the most surprising was the discovery of active volcanism in Io, the Galilean moon closest to Jupiter, which is 262,219 miles from the planet and orbits the planet in 1 day, 18 hours, 27.5 minutes. Scientists speculate that this activity on Io may affect the entire Jovian system.

Voyager revealed Io as a spectacular moon seen as brilliant orange-red with irregular white markings and with a noticeable absence of impact craters. This completely shattered the previously held opinion that it was a cold, lifeless world like our moon, where no change ever occurs. Instead, Io staged a dramatic volcanic display in which eight active volcanoes distributed along the equatorial belt were seen projecting plumes of dark, greenish gas aloft to an altitude in excess of 150 miles. The speeds of the plumes reached 2,200 miles per hour, as against an ejection speed for terrestrial volcanoes of 112 miles per hour. The dark plumes were quite warm—80 degrees F., compared with a surface temperature of 220 degrees below zero F. These plumes are apparently the source of ionized sulfur which circles Jupiter. Those gases which do not escape condense and fall to the surface of Io as a snow—probably sulfur dioxide. Thus, the eruption probably covers the surface of the moon to a depth of perhaps one inch a year. As sulfur is a complex element which can occur in many forms of different colors—reds, yellows, browns, etc.—this provides the vivid colors we see on the moon. Little carbon dioxide and water are found in the plumes, indicating that the moon contains considerably less hydrogen and carbon than the earth. Curiously, the snow that falls is blue. The reason for this unexpected color is that when the gases escape they freeze, and freezing gas forms tiny crystals which scatter blue light and assume this bluish hue. Incidentally, the orbit of Io is not stable. Using positions going back to the seventeenth century, we have determined that its orbit is shrinking at about five inches per year, and the resultant period is shortened by 71 millionths of a second per year.

Jupiter possesses a powerful magnetic field which extends well past Io and creates a huge power station with an output equivalent to that of 1,000 nuclear reactors on earth. As the magnetic lines of force move past Io, the particles which escape from Io react with the lines to create a tube of charged particles linking Jupiter and Io. A 5-million-ampere current has been measured in this "flux" tube.

Why does Io behave in this fashion? Why is it so active volcanically? There is no definitive answer, but some astronomers speculate that from the current generated by the interaction between the Jovian magnetic field and Io, sufficient energy is dissipated to heat the moon's interior by resistive heating—such as is produced by an electric heater. Another hypothesis suggests that tidal imbalances in Io might generate enough friction to heat and melt much of its interior. These imbalances arise when Io's rigid sphere, which is rotationally locked to Jupiter, passes through the gravitational fields of Europa and Ganymede.

Ganymede—the largest of the Jovian satellites—is bigger than Mercury. Its surface, as observed from the Voyager spacecraft, is seen as heavily populated by ancient impact craters. Photographs also disclose bright and dark grooves and ridges criss-crossing the surface, suggesting crustal movements due to tectonic forces similar to those on earth responsible for sea-floor spreading. The occultation of a star by the satellite disclosed the presence of an atmosphere with a density one-hundred-billionth that of the earth. Huge ring basins are seen which show little topographic relief. One large crater is surrounded by a ray system of ejected material. Ganymede, with a density only twice that of water, must be largely composed of water, with a rocky core and an icy crust.

Callisto is also riddled with craters and is probably the most cratered body in the solar system. It possesses an enormous multi-ring meteorite basin that looks almost like a bull's-eye. This has a central, circular patch of light-colored material 180 miles wide. Radiating out from it are eight to ten bright discontinuous but concentric ridges, like ripples in a pond, more or less equally spaced and extending out 800 to 900 miles. Activity following impact events may have produced a viscous flow which tended to flatten out or smooth the surface

features. In composition, Callisto may be composed of equal quantities of rock and ice.

Europa is slightly smaller than our moon. High-resolution photographs by Voyager 2 disclose Europa as a fascinating puzzle, for it appears as one of the smoothest places in the solar system. It has been compared to a billiard ball. It appeared covered by long, shallow criss-crossing dark lines which resemble faults and look as if they had been drawn by a broad pen. The density of Europa is about three times that of water, which indicates that a substantial fraction of the upper crust is ice, perhaps 50 or 60 miles thick; if it is subject to global scale processes, it may have produced the linear systems by moving and shaking Europa's crust. Scientists speculate that the surface may be similar to slushy pack ice found in the Arctic regions.

Amalthea, the innermost moon, is quite red and reflects only about 5 percent of incident light. It apparently is made of dark material, and unlike the other moons, it is irregular in shape. It is 155 miles long by 93 miles wide, with the long axis locked by the gravitational field of Jupiter so that it always keeps the same side toward Jupiter.

As Voyager 1 photographed the dark side of Jupiter, it disclosed several glows which have been identified as lightning bolts several miles long and more powerful than any lightning on earth. Scientists called these "super bolts." These intense lightning bolts serve as efficient low-frequency radio-wave antennae, and the plasma-wave instruments aboard Voyager 1 detected audible "whistler" signals which escape into the magnetosphere. Around the edge of the dark limb of the planet, near the north pole, an aurora 20,000 miles long was seen. Scientists believe it arises from the interactions of electrons with the plasma of ionized particles around Jupiter. These energetic electrons are projected down into the Jovian atmosphere to create the auroral displays.

The fulfillment of a 19-year-old prediction of a ring system by the Soviet astronomer Professor S. K. Vsekhevyatsky of Kiev University was realized when Voyager 1 passed Jupiter. The discovery was a sheer accident—due to the long booms on the spacecraft carrying the magnetometers. These booms are well away from the body of the spacecraft, and as it speeds along, they nod. A photograph of a stellar background was being made. Because of this nodding motion the star trails have little kinks in them, and the long exposure disclosed the spread-out ring system. Thus, Jupiter became the second planet in the solar system known to possess a ring system. These rings, which can be compared to a ribbon around Jupiter, are perhaps a quarter-mile to a half-mile thick. They are approximately 4,000 miles wide, and the inner edge is roughly 37,000 miles from the cloud tops. Photographs from Voyager 2 indicate that additional material of lower density appears to extend the rings all the way in to the cloud tops of Jupiter.

In 1955, Jupiter was discovered to be a strong source of radio energy of various frequencies. It was soon realized that this planet is the second-most-powerful source of radio energy in the solar system; only the sun emits more. These radio emissions occur as intense, sporadic bursts. To duplicate their energy on earth, we would have to detonate million-ton thermonuclear bombs at the rate of one per second.

One mysterious property of these radio bursts is that they appear to be modulated by the position of the satellite Io, the closest of the Galilean moons to Jupiter. There is considerable speculation that the radio emissions are linked to the intense radiation belts around the planet.

SATURN

Even through a small telescope, Saturn, the most distant of the naked-eye planets, is one of the most captivating objects in the entire sky. The reason lies in its beautiful and once unique ring system. When Galileo turned his primitive telescope onto Saturn in 1610, it appeared as a ball with two appendages one-third the size of Saturn. Observations made about two years later startled Galileo, for the appendages seemed to have disappeared, and Galileo concluded that Saturn had indeed devoured its own children. What had happened was that in the first viewing Galileo was seeing the planet with its rings turned so that they might be seen from the earth. Two years later, the rings were edge-on and had disappeared in his low-powered telescopes. It remained for Huygens in 1655 to disclose the true nature of the ring system. With the discoveries of Pioneer 11, apparently hundreds or thousands of rings are circling Saturn, some in tightly-wrapped spiral patterns.

The equatorial diameter of Saturn is 75,000 miles, while the diameter of the readily observable ring system is 171,000 miles. The edge of the newly discovered G ring is 600,000 miles away from Saturn. The innermost edge of the rings (Ring D) is a scant 5,300 miles from the top of the cloudy atmosphere and is not visible in a telescope. The thickness of the rings is variable, with a maximum thickness of about 500 feet. An analysis of the Voyager 1 data revealed that the A ring has a thickness of about 150 feet while the C ring has a thickness of about 30 feet.

Once the ring system was discovered, a tremendous volume of speculation surfaced as to its nature. Theoretical studies indicated that the rings could not be a solid body, and later spectroscopic evidence conclusively proved that the rings were composed of discrete particles of a considerable variety of sizes. Radar measurements made in 1973 indicate that the "bounce-back" signals describe rough, jagged pieces about 3 feet in diameter or larger as the source of these signals. These moonlets cannot be too closely packed together, for stars have been observed through the rings. There is observational evidence that the rings are composed principally of ice crystals mixed, or coated, with other materials of unknown composition. The ice may be the water ice we know on earth mixed with an ammonia ice.

There are at least two divisions in the ring system which are related to the perturbations of the three inner satellites of Saturn. The laws of celestial mechanics prevail, for the entire solar system and indeed for the universe, for just as the planets nearest the sun move more rapidly, so particles making up the inner D ring orbit Saturn faster than those at a greater distance. Particles in the inner ring orbit the planet in about 2 hours, while those in the outer ring circle the planet in 15 hours.

The axis of Saturn is tipped about 27 degrees to its path around the sun. With the year on Saturn approximately 29½ earth years, this means that roughly every 15 years the rings are seen from the earth at their maximum opening, making Saturn a truly magnificent and spectacular object. About seven years later, the rings are turned edge-on to the earth; they disappear in all but the largest telescopes. During the planet's "year" it will thus present the rings wide open twice and edge-on two or three times, depending on the position of the faster-moving earth.

Like Jupiter, Saturn is crossed by cloud belts, which when carefully observed indicate that the planet is spinning on its axis. The period of rotation has been determined by radio signals recorded by Voyager 1 to be 10 hours, 39.9 minutes. This period is precisely controlled by the planet's rotating magnetic field. White roving spots occasionally appear on Saturn and persist for periods on the order of weeks. One spot was observed for a 490-day period between October 1969 and February 1971. The rapid spinning of Saturn results in the planet's being flattened at the poles by about 10 percent.

Telescopic observations disclose that as in the case of Jupiter, the clouds are colored. The equatorial regions are a rather brilliant yellow, and this changes to a greenish hue as the eye sweeps toward the poles. The temperature on the cloud tops is 240 degrees below zero F. At this temperature, most of the ammonia has been frozen out of the atmosphere and may exist as ice crystals. Astronomers believe that hydrogen, helium, argon, and methane must also be present in the thick cloud layers.

There is one intriguing feature of the planet that bears exploring. Because Saturn has, as of early 1982, 21 or 23 satellites, the mass of the planet can be accurately determined. The mass together with the size provides a clue to the density. When this is determined, it is discovered that the density of Saturn is about 0.7—that is, 0.7 times the density of water. This means that if there were an ocean large enough, and we could put the planet into this ocean, the planet would float!

A cross section of the planet reveals that it has an internal structure probably similar to Jupiter's. A theoretical model of the internal structure is, as in the case of Jupiter, modified by the excess radiant energy radiated by Saturn. It appears that Saturn is radiating three and a half times as much energy as it absorbs from the sun. Taking this into account indicates that Saturn has an iron-rock core about 13,000 miles in diameter. This core is surrounded by a 3,000-mile layer of hydrogen ice, which in turn is surrounded by a 5,000-mile layer of metallic hydrogen. Finally, a shell of molecular hydrogen 18,000 miles thick completes the internal structure. The other gases mentioned may be considered contaminants of the hydrogen shell. The outer gaseous atmosphere is less than 200 miles thick. These models of the internal structure of the outer planets are only conditioned speculations, for they are based on assumptions that may bear no relation to reality. As an example, some astronomers believe that almost 70 percent of Saturn's mass is hydrogen. Change the hydrogen content and the characteristics of the internal structure must change.

The first of the satellite family of Saturn—Titan—was discovered by Huygens in 1655. Between 1655 and 1898, eight other satellites were discovered by the observational giants of the astronomical world. However, the tenth satellite, Janus, was discovered by Dr. Audouin Dollfus on December 15, 1966. This addition to Saturn's family is about 300 miles in diameter. The eleventh and twelfth satellites were discovered by Pioneer 11. The eleventh is the satellite closest to the planet and is about the same size as Janus. When an analysis was made of the photographs from Voyager 2, which flew by Saturn in August 1981, additional satellites were discovered. Currently, the estimate is either 21 or 23 satellites. In all probability more will be discovered, for satellites only a few miles in diameter can be detected. The satellites range in size from 200 miles for Phoebe to 3,500 miles for Titan, which means that Titan is larger than the planet Mercury. The large diameter of Titan means that its gravitational field is strong enough to retain an atmosphere, and in 1944 Dr. Gerald P. Kuiper discovered methane in the atmosphere of this satellite.

Studies by Dr. Carl Sagan indicate that Titan has atmospheric conditions quite similar to those on earth. Thus, the giant satellite may be littered with the kind of organic molecules which, in the early history of the earth, led to the origin of life.

URANUS

When Sir William Herschel discovered the planet Uranus in 1781, he believed he had discovered a comet. However, the astronomer Anders Johann Lexell announced in the summer of 1781 that the comet was actually a new planet moving in a fairly eccentric orbit at about nineteen times the earth's distance from the sun. When the records of old observations were examined, it was found that it had been seen as early as 1690 by the first Astronomer Royal, John Flamsteed, as "a star of the sixth magnitude." Between 1690 and 1781, almost a hundred years later, many astronomers had observed the planet but failed to recognize its true nature. The eccentricity of the orbit is such that the difference between its closest approach to the sun and its greatest distance from the sun is 84 million miles. This variation in distance is within 9 million miles of the earth-sun distance. Curiously, Uranus is of the sixth magnitude and can just be seen with the naked eye under the most ideal observing conditions. However, as a year on Uranus is 84 earth years in length, the planet moves very slowly, and while the ancients undoubtedly saw it innumerable times, the slowness of its motion and lack of movement against the background of the stars did not excite curiosity, so it remained undiscovered until the eighteenth century.

As is characteristic of the outer planets, Uranus is covered by cloud belts. It is so far from the sun that the temperature of the tops of the cloud belts we see must be 300 degrees below zero F., or even colder. It has an aquamarine hue, which indicates methane in the atmosphere. Actually, the spectrum of the planet shows bands in the red, orange, and green that are similar to those we find in the spectra of Jupiter and Saturn. For this reason, astronomers speculate that the dense clouds contain significant amounts of ammonia in the form of crystal. These cloud belts permit the planet to reflect only about 50 percent of the light falling on it. The velocity of the clouds on Uranus has been measured at over 200 miles per hour, compared to about 85 miles per hour for terrestrial clouds.

Like the earth, Uranus has a magnetic field which is rakishly tilted to its spin axis by about 60 degrees, compared to about 11 degrees for the earth. However, the strength of the field is only slightly less than that of the earth's. Astronomers speculating on the reason suggest that Uranus is a three-layered planet with a central core of heavy material composed of iron and rock. Over this is a mantle of water, ammonia and methane several thousands of miles deep. Surrounding this is an atmosphere composed primarily of hydrogen and helium several thousand miles thick.

The most startling and bizarre feature of Uranus is that it orbits the sun lying almost on its spin axis. This spin axis, like the north pole of the earth, is always pointed to a particular point in the sky. As a matter of fact, if we continue to think of the planets rotating in the same direction as that of their revolution around the sun (this, as we have discovered, is not true in the case of Venus), we could say that the axis of Uranus is tipped over 98 degrees to the plane of its orbit.

The revolution around the sun with its axis always

pointed to the same part of the sky would provide to a terrestrial observer a unique look at the planet. If we observed Uranus at a particular moment, we might be looking down on it from above one of the poles. One quarter of a revolution later—twenty-one earth years later—we would be viewing the planet from above the equator. Another twenty-one earth years later we would be looking down on it from a point above the other pole. No other planet in the solar system behaves in this fashion.

The day on Uranus indicates that, like Jupiter and Saturn, it too rotates on its axis very rapidly. The most recent determination of the planet's rotation indicates that the day has a period of about 17.3 earth hours. With this rapid rotation one would expect that Uranus is, and indeed it is seen as, oblate, with the equatorial diameter considerably larger than its polar diameter.

Uranus possesses many satellites. They range in size from the 150-mile diameter of Miranda, discovered by Gerard P. Kuiper in 1948, to the 1,100-mile diameter of Oberon and Titania, named for characters in Shakespeare's plays. The other previously known satellites, named for characters in Pope's "The Rape of the Lock," are Ariel and Umbriel. Voyager 2 disclosed ten more moons circling Uranus, for a total of fifteen. The newly discovered ten are so small that no names have been assigned to them. There are probably many more satellites five miles in diameter or smaller that circle Uranus, but the resolution of the photographic systems on Voyager 2 was unable to discern them. The major moons are composed largely of ice, intermixed with rock and methane frost. They are all surprisingly dingy, reflecting only 20 percent of the incident light.

Observations on March 10, 1977, with the 36-inch telescope aboard the high-flying Kuiper Astronomical Observatory, disclosed that Uranus has at least nine spider web–thin rings, composed of extremely dark material, circling it in its equatorial plane, which form a ring system similar to that possessed by Saturn. This was confirmed by Voyager 2, which disclosed two other distinct rings. The rings, together with the many satellites, trace out a target-like bull's-eye. Some fragments of rings were also discovered, and those studying the ring system believe there may be as many as one hundred rings and fragments of rings around Uranus.

Curiously, this discovery fulfilled a prediction made in 1973 by Gordon Taylor of the Royal Greenwich Observatory, who indicated that Uranus would occult a ninth-magnitude star, SAO 158687, on March 10, 1977, and that this would be the ideal opportunity to determine the presence of a ring system around Uranus if one existed. When the dip in the light intensity was first seen, astronomers believed that this might be due to the presence of a

satellite. It was only after looking at the complete set of photometric tracings that James Elliot, leader of the Cornell University group, saw the preimmersion (inbound) and postimmersion (outbound) occultations which indicated that indeed a discrete ring system circled Uranus.

The rings must be quite thin, for the star was never obscured completely. Because of this, astronomers speculate that the rings are composed of fairly large-sized particles. These are very dark, reflecting light very poorly. This lack of reflectivity accounts for the extreme difficulty of seeing or photographing the rings. The rings are contained in a fairly narrow belt lying about 11,000 miles above the cloud tops of Uranus and extending out to perhaps 65,000 miles. The first four are really ribbons about 8 miles wide and almost circular. The outermost may be up to 60 miles wide and may be not circular but of variable width. As in the case of Saturn, the gaps in the rings are probably due to resonance effects which dictate the periods of the particles in the ring as precise fractions of the periods of the satellites. There has been considerable speculation as to the coal-black color of the rings; some astronomers suggest that the intense radiation belts around the planet transform methane into a black amorphous polymer, which accounts for their extraordinarily low reflectivity.

NEPTUNE

When the seventh planet, Uranus, was discovered, an orbit had to be computed for it. Astronomers in their computations take into account the perturbations, or gravitational influences, of all the other planets. When these perturbations are accounted for, it should be possible to compute an accurate orbit for the planet so that its future positions may be precisely determined. By 1788, the tables computed for Uranus failed to represent the planet's position with acceptable astronomical accuracy. By 1821 Alexis Bouvard came to the conclusion that ancient and modern observations of Uranus could not be reconciled, and he concluded that this was due to some "foreign and unperceived cause" which might have been exerting an influence on the planet. However, beyond making this perceptive comment, Bouvard could not do a thing. In another few years, some astronomers were freely speculating on the possibility of a trans-Uranian planet.

In July 1841, John Couch Adams, at the young age of twenty-two, decided that as soon as possible after receiving his degree he would try to determine the elements of the orbit and the physical characteristics of an undiscovered planet lying beyond Uranus. He began his computations in 1842 and by 1845 had determined the orbit of the eighth planet. The positions he provided were within two degrees (four moon diameters) of its true position, and if an attempt had been made, the eighth planet would have been discovered in October 1845.

Across the English Channel, in France, another brilliant mathematician, Urbain J. J. Leverrier, tackled the problem of the irregularities in the orbit of Uranus. His preliminary work indicated that these irregularities increased with time. He persisted in his efforts to account for these and finally, early in 1846, arrived at the conclusion that only the presence of a planet beyond Uranus could explain them. By August 1846 he not only had computed the position of the unknown planet but had indicated that it should show a disk to distinguish it from a star. On the evening of September 23, 1846, the German astronomer Johann Galle in Berlin, in a response to the request of Leverrier to look for the planet, found it within an hour after beginning his search in substantially the identical position deduced by Adams some eight months earlier. It was a disk 3 seconds of arc in diameter and of the eighth magnitude within 55 minutes of arc (less than two full moon diameters) of its predicted position.

With the discovery, James Challis, the English astronomer, found that he had observed the newly discovered planet four times without realizing the fact. Thus, priority for the discovery went to Leverrier, though Adams had been the first to solve the problem. Later, when the bitterness of nationalism had evaporated, Adams and Leverrier were considered codiscoverers of Neptune. The planet was named Neptune by Leverrier, and a trident was accepted as its symbol. Thus came into our realm of knowledge the first planet whose existence and position had been predicted theoretically.

As Neptune is of the eighth magnitude, it can be seen with a small telescope or binoculars and presents a disk with a greenish cast. Because of this, one can assume that the surface clouds we see have characteristics similar to those of the other three major planets. In all probability, the greenish color arises from the presence of methane in its atmosphere. Neptune also has an albedo of 50 percent.

The planet is so far from the sun that it takes 165 earth years to swing around it once. Recent measurements indicate that the day is about 18.2 earth hours long. The short day and the continuous precession also indicate that the planet is not round; its equatorial diameter must be somewhat larger than that at the poles.

Very little is known about the internal structure of either Uranus or Neptune. Like Jupiter and Saturn, they are believed to contain large amounts of lighter elements. However, astronomers can construct theoretical models, even though it is unknown whether these two planets have hot or cold interiors or whether they have a solid surface. Astronomers speculate that these planets have a rocky core 9,000 to 10,000 miles in diameter. This is covered by an icy

mantle about 5,000 miles thick, and the mantle is surmounted primarily by molecular hydrogen. Above the hydrogen layer are ammonia and methane.

There are two satellites in Neptune's family. The largest one, Triton, was discovered by the English astronomer William Lassell on October 10, 1846, and a smaller one, Nereid, was discovered by Kuiper in 1949. Triton must be somewhat larger than the earth's moon and moves around the planet in a circular orbit. Nereid, on the other hand, moves around Neptune in a highly eccentric orbit, varying its distance from 900,000 to 6 million miles. Its estimated diameter is less than 200 miles, which may be one reason it was not discovered until 1949. Curiously, its period—that is, the time it requires to circle Neptune—is about one earth year: 360 days, to be precise. Triton has an atmosphere of methane and nitrogen covering an ocean of liquid nitrogen. It may even have solid methane "icebergs" floating in the nitrogen ocean.

In 1981, critical observations indicated that perhaps a third satellite was orbiting Neptune. Later observations of star occultations led astronomers to conclude that, instead of a satellite, there are at least two arcs or parts of rings around Neptune. The arcs were detected by watching a star suddenly dim as it passed behind an arc. They estimate that the arcs block fifteen to forty percent of the star's light and, from the duration of the dimming, they calculate that the arcs must be from ten to twenty miles wide. The arcs are apparently in the equatorial plane, and they are spaced 32 and 50 miles above the atmosphere. If these are indeed found to be arcs, then they are unlike anything else in the solar system, for they violate the theory which mandates complete ring systems. We may have to wait until August 1989 when Voyager 2 visits Neptune to resolve the mystery surrounding the arcs.

While Pluto is normally the most distant of the planets, on January 22, 1976, it moved inside the orbit of Neptune, bringing it closer to the sun. Thus from January 1976 to March 1999, Neptune is the farthest planet from the sun.

PLUTO

Pluto, the outermost and ninth planet in the solar system, was also discovered as a consequence of irregularities in the orbit of Uranus. The first discrepancy to surface was that the mass of Neptune had two values: one was derived from the perturbations it produced in the orbit of Uranus; the second arose from the motion of Neptune's satellite Triton. However, if a trans-Neptunian planet did exist, its position would have to be determined from the irregularities in the orbit of Uranus and not those involving Neptune. By 1879, the French astronomer Camille Flammarion believed that a planet external to Neptune might be revealed by its actions on comets which, at their greatest distance from the sun, would move into the gravitational domain of the unknown planet. By 1900 there were published predictions of the existence of this remote planet.

The most significant work on determining the reality of Pluto was done in the United States. In 1909, Professor William H. Pickering published a paper on the search for a planet beyond Neptune, and Percival Lowell published "Memoir on a Trans-Neptunian Planet" in 1915. Professor Pickering chose to use a graphic method to determine the position, while Lowell employed a rigorous computational method.

Lowell, in the year 1905, began a photographic search that lasted until 1907. Lowell's plates uncovered hundreds of nonstellar celestial objects such as planetoids, comets, flaws in the plate, and even variable stars, but no ninth planet. Seven years later, in 1914, a second search was begun at the Lowell Observatory. Percival Lowell died in 1916, and with his death the second search terminated. In 1919 a third search, based on the work of Professor Pickering, was organized at the Mt. Wilson Observatory. This too proved futile. Ten years later, a fourth search was begun at the Lowell Observatory, but this time a new and highly sophisticated piece of equipment, the blink comparator, was used. This was a precisely made machine that permitted the astronomer to scan two accurately positioned plates alternately, star by star. If a star appeared on one and not the other, then it was a sign that one of the stars had shifted its position, and this warranted further investigation.

During late January 1930, Clyde W. Tombaugh photographed a region of the sky near the star Delta in the constellation of Gemini, the Twins. On February 18, Tombaugh caught a yellowish object of the fifteenth magnitude popping into and out of the background. The director was notified, and further checking went on. Finally, on March 13, 1930, the decision was made to announce the discovery of the trans-Neptunian planet. (On this day 149 years earlier, Herschel had discovered Uranus.) The planet was named Pluto, because Jupiter and Neptune were that god's mythological brothers, and the symbol was an intertwined P and L or ♇, which were Percival Lowell's intitials.

Pluto is very faint; it is in fact small, with a diameter and mass less than those of the earth. But in his computations Lowell had assumed a mass considerably larger than the earth's. The true mass of Pluto was too small to produce the perturbations on Uranus on which Lowell had performed his computations. Yet the planet was found within 6 degrees of Lowell's prediction! This could well be one of the most fortuitous coincidences in the history of astronomy.

The orbit of Pluto is the most eccentric in the solar system. At perihelion, its closest approach to the sun, it is 2.78 billion miles from that fiery body. At aphelion, it is 4.5 billion miles from the sun. This means that at perihelion it is closer to the sun than Neptune. It moved in closer to the sun in January 1976, and will remain so until March 1999. Could there be a collision of Pluto and Neptune when Pluto is within Neptune's orbit? The answer is no. The reason is the 17-degree inclination of Pluto's orbit to the ecliptic. The closest approach of these two planets will be about 240 million miles.

Drs. Merle F. Walker and Robert Hardie had found that Pluto varies in brightness by about 10 percent, with a regular period of 6.39 earth days. This variation has been attributed to rotation. This is a long period of rotation for a planet. Could this mean that Pluto was once a satellite? Dr. Kuiper believed that this was the case and that in some as yet unexplained fashion Pluto was torn loose from Neptune to be sent orbiting the sun; this, by definition, makes Pluto a bona fide planet.

On June 22, 1978, Dr. James W. Christy of the United States Naval Observatory discovered a moon circling Pluto. He had noted an elongation of the photographic image of Pluto, and when he examined photographic plates taken in 1965 and 1970 he discovered similar elongations, always in a north-south direction relative to the earth. The only plausible explanation for this stretching is a moon with a diameter of somewhat less than 1,000 miles circling Pluto at an altitude of about 8,500 miles. Orbital elements derived by Dr. Robert S. Harrington indicate that the moon circles Pluto in an almost circular orbit in about 6.4 earth days, which is the same as the Plutonian day. This means that the moon, named Charon, for the mythological boatman who ferried souls across the River Styx into the underworld, hangs stationary over the same spot on Pluto's surface, behaving in the same fashion as a synchronous satellite of the earth. However, because Charon's diameter is half that of Pluto, many astroners are considering Pluto to be a double planet instead of a planet with an extraordinarily large moon.

Pluto is so distant from the sun that it takes 248 earth years for a single orbit. Because of this distance, the temperature is lower than 373 degrees below zero F. At this temperature, the gases, probably ammonia or methane, appear to be frozen out of its atmosphere as a layer of ice. Infrared observations verified the presence of methane ice in 1976. There is the possibility that hydrogen, helium, and neon might be present, but it is not possible to detect their presence spectroscopically.

Pluto's physical characteristics are almost impossible to determine. Astronomers had assumed a diameter of about 3,600 miles. More recent figures provide a diameter of about 1,800 miles and a mass of 0.002 of the earth's mass, which would give Pluto a density less than that of water—0.7 is the probable value. This means that Pluto is nothing more than a giant snowball of frozen gases.

DR. I. M. LEVITT
April 1986

METEORS:
"HOT RODS" OF THE SOLAR SYSTEM

By watching, I know that the stars are not going to last.
I have seen some of the best ones melt and run down the sky.
Eve's Diary, *Mark Twain*

IF YOU ARE outdoors tonight after midnight and watch the clear dark sky for fifteen to twenty minutes, you will probably see at least one "falling" or "shooting star." This faint light streak slashing across the night sky is a meteor. If it is spectacularly bright, it may arouse your curiosity about what causes shooting stars, where they come from, and how frequently they appear.

Meteors come from interplanetary space and burn up in our atmosphere. In space, they are called meteoroids, and most are not larger than a pinhead. They pursue a path under the influence of the sun's gravitational pull with a velocity of between 18 and 26 miles a second. Their space temperature is low, just a few degrees above the melting point of ice because of their exposure to sunlight. How do these tiny bits of star stuff give rise to a superheated, glowing column of air several miles long and several feet in diameter? The answer lies in their swift motion.

When these tiny grains collide with the earth's atmosphere, the friction on the particles as the molecules of air bombard their forward surface heats the outer shell to a temperature of from 3000 to 4000 degrees Fahrenheit, and the material begins to flow and evaporate. The resulting vapor particles collide with billions of molecules of air at high velocities; the energy of the collisions is absorbed and then released as radiation. Meanwhile, the material not yet vaporized is continually flowing in liquid form over the front and sides of the body. Eventually, small meteoroids are completely vaporized in flight and vanish after burning out. That tiny luminous streak that we observe represents the demise of a tiny particle that entered the earth's atmosphere.

Meteors are far more common than people suspect. On average, we may see approximately eight meteors per hour. After midnight, the count may mount to a dozen or more per hour. The increase is due to the earth's motion. As the earth rotates in the same direction as it orbits the sun, between noon and midnight we are on the rearward side of the earth, and we are looking in the direction from which the earth has come. From midnight until noon, we are on the forward side of the earth; we are moving in the direction of the earth's travel. This means that between noon and midnight the meteoroids are catching up with the earth and the closing velocity is zero.

When the earth meets the meteoroid head-on, the velocity of the earth around the sun (about 18 miles per second) is added to the velocity of the meteoroid (about 26 miles a second), resulting in a collision speed of about 44 miles a second!

At times, meteors put on such a pyrotechnic display as to deserve a place in history. One such shower took place on November 12, 1833. According to reports in the *Georgia Courier:*

> At about 9:00 P.M., the shooting stars first arrested our attention, increasing in both number and brilliancy until 2:30 A.M., when one of the most splendid sights perhaps that mortal eyes have ever beheld was opened to our astonished gaze. From the last-mentioned hour until daylight the appearance of the heavens was awfully sublime. It would seem as if worlds upon worlds from the infinity of space were rushing like a whirlwind to our globe . . . and the stars descended like a snowfall to earth.

In Boston that night, one actual count for the fifteen minutes before 6:00 A.M. came to a total of more than 30,000 meteors per hour!

While it was this shower in 1833 that produced a burst of interest in the evanescent objects, there had been one of equal importance on November 11, 1799. Showers were seen on November 13, 1866, in Europe, and one year later in America. These meteors all appeared to emanate from a small region in the constellation Leo, thus named Leonids. In 1901, the rate of the Leonids was only about 800 per hour, and since then the shower has been disappointing.

In the second week of August each year, there is a fairly satisfying shower radiating from the constellation Perseus; a good average rate at the peak of the shower might be fifty to sixty per hour. Like most meteor showers, the Perseids gain their name from the constellation from which they emanate. The Perseids are sometimes called by another name given them by the Irish peasants—the "tears of St. Lawrence," his feast day falling on August 10. This is perhaps the best known of the approximately ten meteor showers that visit us each year. There is a shower from Orion in middle October, another from Gemini in mid-December, and other showers of less importance throughout the rest of the year.

The mystery of where meteors come from was solved in 1866, when the Italian astronomer Giovanni Schiaparelli computed an orbit for the Perseid meteoroids. The period he assigned was 108 years. The characteristics of the orbit coincided closely with the orbit of the comet discovered July 15, 1862, by Lewis Swift. The period of revolution of that comet was computed to be 122 years, against the 108-year period Schiaparelli assigned to the Perseids, but this agreement is considered good. Schiaparelli felt no hesitation in announcing that the bodies that produced the Perseids were traveling in a path identical to that of Swift's comet of 1862.

Comet Tempel (1866) and the Leonids have the same path, with a period of 33 years but the Leonids have almost disappeared, as though their paths had been altered. The Lyrids of April pursue the orbit of Thatcher's comet (1861); the Eta Aquarid meteors and the Orionids are strewn along the path of Halley's comet, which last appeared in 1985; the Taurids follow the orbit of Comet Encke, with the shortest known period, 3.3 years. The Giacobinids thrust themselves into prominence on October 9, 1933. As darkness descended on Europe, a great meteor shower was already in progress. From the head of the Draco constellations, meteors were streaming at the rate of 20,000 per hour for a single observer! The comet with which these meteors are associated was discovered by Giacobini in 1900.

Because of these associations of comets and meteor streams, most astronomers believe that most meteoroids arise from the solid portions of comets. We know that most of the comet is practically nothing at all. There is a loosely bound mass of gases and other finely divided material we might call dust in the center of its brightest portion—the nucleus—which probably consists of a swarm of meteoroids. When a comet is far from the sun, only the nucleus and a roughly spherical cloud of surrounding gases are visible. As the comet comes closer to the sun, the radiation boils other gases from the meteoroids of the nucleus, to enlarge the head of the comet and to produce its far-flung tail. But the meteoroids are only loosely bound by their mutual gravitational attractions, and disturbances of the planets can separate individuals from the swarm to scatter them all along the path of the comet. It is these particles that make up the swarms that produce meteor showers. These showers come at predictable times, for they lie in the orbit of the comet that gave rise to them. However, at certain times the showers put on a truly magnificent display like the one in 1833. These superb displays occur because of the nature of the comet and its orbit.

As indicated, the comet leaks meteoroids, which in time outline its orbit. If a comet has been leaking particles for a long time, the orbit of the comet will have been gradually filled with particles. Once a year, when the meteor-strewn orbit of the comet crosses the earth's path around the sun, we see a small shower of meteors (indicated in the chart opposite). But when the concentrated *nucleus* of the comet intersects the earth's orbit, which occurs once in the comet's periodic orbit, then the nominal shower becomes a veritable rain of meteors. Thus, in the case of Swift's comet, with a period of 105 years, the nucleus intersects the earth's orbit once every 105 years. That is what occurred in 1862, when a strong shower was seen. In the case of Tempel's comet, the nucleus intersected the earth's orbit in 1866, and again an intense shower was seen.

In addition to rather pedestrian meteors, there is another class of "hot rods" that will momentarily flash into view and then disappear, leaving a long, bright streak and perhaps a trail. They are fireballs, which may be traced back to a shattered asteroid. Fireballs may surpass the moon in brilliance and even cast shadows. No larger than your thumb, they are capable of causing mild panic among their observers. One working definition of a fireball is that it is a meteor so bright that people will talk about it. We get only five or six fireballs per day for the whole earth, so the probability of seeing one is rather low.

The average fireball begins to glow at a height of 65 miles and travels at a speed of about 20 miles per second. It will traverse a slanting path about 80 miles long in a few seconds. On rare occasions, it will be seen to explode into discrete pieces near the end of its path. These pieces are called "bolides." Some of them survive their passage through the atmosphere, ending their careers in a momentary blaze of glory and falling to the earth where they become meteorites. Occasionally, they are picked up as pieces of iron, stony-irons, or stones. These are the only things that did not originate on the earth that one can touch. They are literally pieces of "star dust."

In 1807, President Thomas Jefferson heard of a stone that plummeted to earth at Weston, Connecticut. He supposedly grumbled, "I could more easily believe that two Yankee professors would lie than that stones would fall from heaven." Currently there can no longer be any doubt that things do fall from the sky, and, when found, are put into museums where we can marvel at these celestial vagabonds.

DR. I. M. LEVITT
February 1992